HOSHIN* PLANNING
THE DEVELOPMENTAL APPROACH

Organizational Hierarchy of Needs

BY BOB KING

© Bob King, GOAL/QPC
1989

ISBN 1-879364-00-X

*Based on the Japanese system hoshin kanri, hoshin planning is also called policy deployment, management by policy (MBP), and management by planning.

Table of Contents

Preface i

Chapter 1: Hoshin Planning and Total Quality Management 1-1
    Part 1: The Nature of Hoshin Planning 1-1
    Part 2: Hoshin's Role in Total Quality Management 1-3
    Part 3: Key Aspects of TQM Related to Hoshin Planning 1-12
    Part 4: Daily Control and Hoshin Planning 1-16

Chapter 2: Planning Problems and Hoshin Solutions 2-1
    Part 1: Some Problems with Planning 2-1
    Part 2: The Principles of Hoshin 2-3
    Part 3: The Hoshin Planning System 2-8

Chapter 3: Problem Solving Tools 3-1

Chapter 4: The Seven Management Tools 4-1
    Affinity Diagram/KJ Method 4-2
    Interrelationship Digraph 4-6
    System Flow/Tree Diagram 4-10
    Matrix Diagram 4-15
    Matrix Data Analysis 4-26
    Process Decision Program Chart (PDPC) 4-28
    Arrow Diagram 4-33

Chapter 5: Target/Measures Matrix 5-1

Chapter 6: The Flag System: One Tool for Alignment 6-1

Chapter 7: Other Tools for Alignment 7-1

Chapter 8: Hoshin Planning Phase 1: Process Management 8-1
    Part 1: The Developmental Need for Process Management 8-1
    Part 2: Tru-Save Example 8-4
    Part 3: Dr. Deming's 14 Points 8-11

Chapter 9: Hoshin Planning, Phase 2: Management Self-Diagnosis 9-1
    Part 1: The Developmental Need for Self-Diagnosis 9-1
    Part 2: Example 9-4

Chapter 10: Hoshin Planning, Phase 3: Alignment of Targets and Means with Long Range Vision and 1 Year Plan 10-1

Chapter 11: Advanced Hoshin Planning 11-1

Appendix A

Appendix B

Index

Glossary

Conclusions

Bibliography

Preface

In 1950, the Japanese established the Deming Prize to recognize companies who changed from traditional, ineffective management to the successful management style known as Total Quality Management.[1] Companies who won this award in the 1970's and '80's reported that hoshin planning played a central role in their management. Recounting a couple of these successes may indicate the magnitude of this subject.

Yokagawa Hewlett-Packard, a division of Hewlett-Packard, and Kayaba, the leading supplier of shock absorbers to the Japanese auto industry, have demonstrated how hoshin planning can lead to great progress -- even the coveted Deming Prize.

Yokagawa Hewlett-Packard

In the early 1970's, Yokagawa Hewlett-Packard (YHP) was one of the least successful of Hewlett-Packard's divisions. YHP's quality was poor, the sales were low, and the profit was low. The people at YHP decided to implement Total Quality Management. They used hoshin planning to orchestrate the transformation involved. By the late 1970's, much progress had been made. YHP had high quality, good market penetration, and good profit. They were now the leading division in Hewlett-Packard in these indicators and have maintained that position during the 1980's by continuous improvement orchestrated by hoshin planning.

Initially YHP made improvements in individual areas. The reduction of defects in dip soldering is an example. (See Figures P.1 and P.2) The results are particularly impressive considering that this was accomplished with old, unwanted equipment which YHP had received from other divisions.

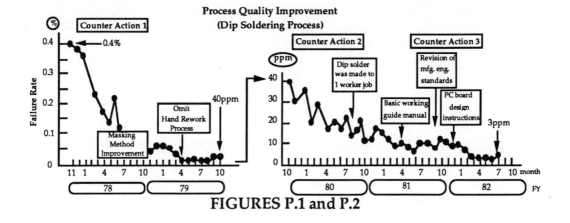

FIGURES P.1 and P.2

---

[1] The Japanese translation is usually either Total Quality Control or Company-Wide Quality Control. This text uses "Total Quality Management" because it is a translation which is more popular in the U.S.

As more projects were accomplished the results were even more dramatic. Some examples include:

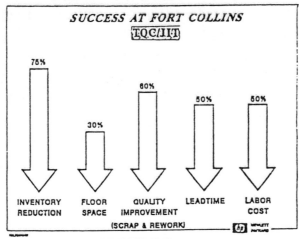

FIGURE P.3

In the early 1980's Hewlett-Packard had committed to Total Quality. Company president John Young had challenged the corporation to ten-fold improvements in key indicators during the decade of the 1980's. Hewlett-Packard employees responded enthusiastically; statisticians pooled course materials and developed new ones. Employees went to work on key projects and had many successes. By 1985, many key indicators had improved by a factor of five. It became clear in 1985 however, that the challenge of a ten-fold improvement would not be reached at their current progress because it had taken five years for the first five-fold improvement and the second five-fold improvement would be more difficult.

YHP continued to make astonishing progress. In 1985, Hewlett-Packard executives asked YHP what was needed for further improvement. YHP responded that hoshin planning was the fundamental missing system. In 1986-'87, Hewlett-Packard implemented the model of hoshin planning that YHP had adopted in the 1970's in the quality area. In 1987-1988 they implemented it throughout the company. It is still too early to tell whether this will be enough to help YHP reach their 1990 goals.

Kayaba

Kayaba is the leading manufacturer of shock absorbers for the automotive industry in Japan. In 1973, as a result of pressures from the oil shock, Kayaba informed customers that it was raising its prices. The customers' response was that they could not absorb the price increase and that Kayaba's quality which was once quite good, was deteriorating. These comments got Kayaba's attention and they launched an investigation to get further information.

Kayaba learned that their quality had truly declined. They also decided that they had been simply filling customer orders rather than anticipating them. They needed to develop a system to anticipate customer orders. They also needed to conduct research of their own rather than wait for the research of their customers so that they could store their new technology and release one product after another just as customers were ready for them.

These improvement efforts on quality and anticipatory development (later called Quality Function Deployment) were orchestrated by hoshin planning. In the quality area Kayaba asked each employee what went wrong and what they worried might sometime go wrong. They developed a plan and a five-year gant chart to indicate when they would work on these problems. The following is a partial example.

Step	Problems prior to introduction of CWQC	Activities	Starting from '76 '77 '78 '78 '79	Division responsible	Documents Guidelines and standards
Product planning	1. Insufficient grasp and analysis of market requirements	- Obtain accurate grasp of market quality requirements by strengthening system for gathering technical data		Engineering division	Guidelines for new product development Data collection manual Quality table manual Production engineering manual
	2. Unsuitable goals for quality and engineering	- Establish goals anticipating market requirements. - Highlight correlation between quality and engineering.			
	3. Inadequate correlation of quality and engineering with cost	- Establish cost control system meshing with QA system and use it to implement cost planning		Cost control division	Cost control guidelines
Product design	1. Inadequate model for basic engineering	- Improve theoretical analysis of engineering. - Improve correlation of theoretical and experimental work.		Engineering division	Quality table manual Production engineering manual
	2. Insufficient quality deployment and evaluation	- Make use of quality deployment techniques. - Improve FMEA and OR. - Develop methods for evaluating that products under outdoor analogous to those of actual use.		Engineering division	OR implementation
Production preparation	1. Problems occurring in the introduction of new products	- Improve process planning. - Introduce and use FMEA for processes.		Production engineering division	Production engineering manual Process FMEA manual
	2. Insufficient quality evaluation of initial production lot	- Improve control and evaluation of initial production lots.		Manufacturing division, QA division	Control guidelines for initial production lot Inspection planning manual
Production	1. Lack of systematic process control on account of emphasis on quantity over quality	- Review process control. - Improve control of key processes. - Improve quality recording.		Manufacturing division	Control guidelines for key processes Problem-prevention system diagram
	2. Frequent recurrence of complaints on account of weak response to complaints and process defects	- Promote measures for preventing problems. - Improve control of irregularities. - Furnish QC guidance to key supplier plants.		Manufacturing division, purchasing division	Guidelines for irregularity control QA essentials for customers
	3. Unaggressive approach to study and analysis of process capability	- Step up efforts on behalf of improving equipment. - Study process capability. - Do more analysis of processes.		Manufacturing division	Study manual for process capability
	4. Inadequate inspection system leading to problems for customers	- Clearly define authority for suspending shipment. - Carry out promised inspection. - Improve trial manufacture for confirmation.		QA division	Rules for shipment suspension Essentials of inspection standardization
Sales and service	1. Insufficient measures for preventing recurrence of customer complaints	- Improve surveys of product drawing complaints. - Strengthen analysis and feedback on complaints. - Report major quality problems to facilitate their solution. - Promote analysis of product failures.		QA division, marketing division	Guidelines for processing major quality problems Standards for headquarters registration of quality problems
	2. Analysis of potential for failure insufficient and not linked to efforts preventing complaints				
Comprehensive monitoring	1. Insufficient performance evaluation and feedback in regard to quality targets	- Define responsibility and authority for quality evaluation and inspection. - Horizontally deploy aspects of successful QC work (supervised by joint evaluation committee). - Monitor quality.		QA division	System diagram for quality evaluation and product inspection Essentials of work by joint evaluation committee Essentials of quality monitoring
	2. Monitoring insufficient and not linked to system improvement				

FIGURE P.4

They also worked on a system chart for anticipatory development. This is an example:

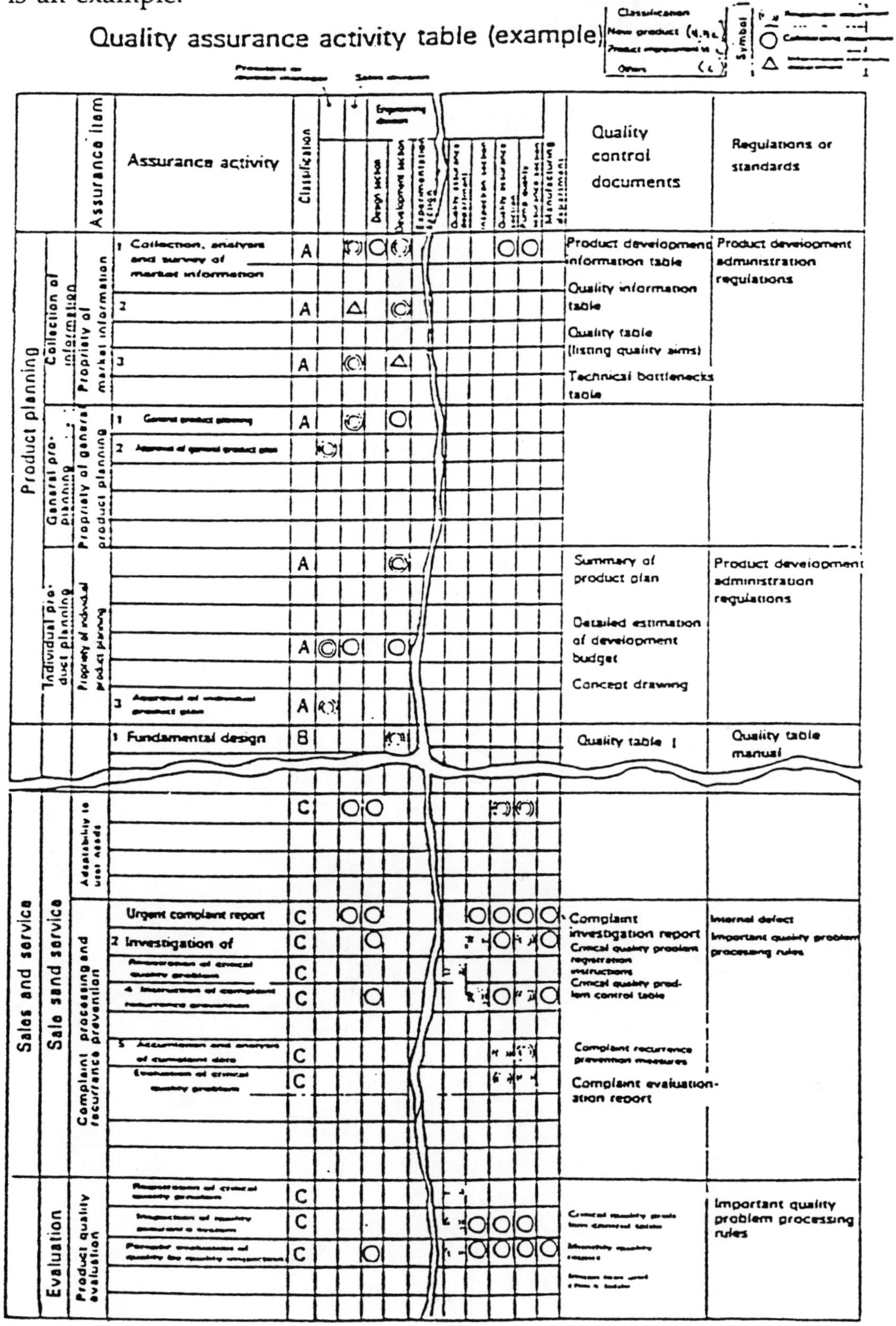

Quality assurance activity table (example)

FIGURE P.5

These two examples, along with others, have been an incentive to write this book. Hopefully this will serve as a basis for developing a great, long lasting competitive advantage for your company.

Chapter 1
Hoshin Planning and Total Quality Management

During 1980 and 1981, the author and members of the GOAL/QPC staff began studying Dr. W. Edwards Deming. In 1981 and 1982, that study grew to include Philip Crosby and Joseph Juran. Each new learning expanded their horizons.

But when the GOAL/QPC staff went to Japan in 1983 to study Japanese quality experts, they learned a rather shocking insight. The Japanese had not only copied the United States beginning in 1950 with Deming and in 1954 with Juran, but they had also surpassed the U.S. in certain quality and management systems. An even bigger surprise was that some of these new and important Japanese systems were basically unknown in the States. They were unknown and unused by companies who had sent dozens, maybe even hundreds of executives to Japan, unknown by the leading business schools in the country, and not taught in any MBA programs with which the author is familiar.

One of those new systems was hoshin kanri or hoshin planning. This introductory chapter will focus on the nature of hoshin planning and its role as part of Total Quality Management, with a special focus on hoshin's relationship with a customer driven master plan and daily control.

PART 1: The Nature of Hoshin Planning

Part 1 includes obstacles to learning hoshin, hoshin's role, and its name.

Barriers to Learning Hoshin Planning

There are a number of reasons why these Japanese systems were not used in the States. One of the reasons is that many of the Japanese believe that these systems will not work in the States because of cultural differences and therefore don't describe them to their American visitors.

A more important reason is that the Japanese have a different philosophy of education than we do. Their philosophy is that when the student is ready, the teacher will appear. What that means is that you should teach someone the next thing that he should learn in a sequence of learning. For example, before you go on a tour in Japan, the Japanese want to know how much you know about their country, so that they know what piece to present to you next.

Another reason Americans fail to learn these systems is that the Japanese teach by puzzles. Starting in kindergarten they give students puzzles to solve. As the student solves the puzzle, he learns. So even if you translate the leading Japanese quality management text, you end up with an English-Japanese puzzle which needs to be solved to be understood, and that is not an easy task.

Hoshin, A Foundation System in Japan

In October of 1988, Michael Brassard of the GOAL/QPC staff visited half a dozen Japanese companies and asked each of them about various Quality Management Tools and Systems. Hoshin kanri is the only system from the list that was known and used by every company interviewed. Hoshin kanri, translated as "policy deployment", is one of the core aspects of Japan's management system. Kenzo Sasaoka, the President of Yokagawa Hewlett-Packard which is the Japanese Division of HP, said at the GOAL/QPC conference in November of 1988 that hoshin planning is the foundation of Hewlett-Packard's quality management effort.

What's in a Name?

Hoshin kanri is the Japanese name for hoshin planning. In Japanese, these words mean "shining metal" and "pointing direction". Hoshin planning is a system that points the organization in the right direction. The more common translation of hoshin kanri is policy deployment.

When Deming and Juran went to Japan in the 1950's, the Japanese didn't want to translate the English quality terms into Japanese because they were afraid they would lose something in the translation. So if you read a Japanese quality manual you may suddenly see things like statistical quality control, run chart, and histogram among the Japanese characters. When the Japanese began to develop new systems, they thought the words ought to be in English, which was very unfortunate for us, because the English they used was "Japanese-English" which is usually either awkward or even misunderstood by the American reader.

Take policy deployment, for instance. What does policy deployment mean? If you tell people in your organization that you are studying policy deployment they will probably think you are studying personnel policies. Most people guess that policy deployment is a system for making sure every employee behaves according to company personnel policies. That is not what it is at all.

Chapter 1: Hoshin Planning and TQM   p.1-3

The word hoshin can be translated as "policy" or "target and means". The word kanri is translated as "planning" but also means "management" and "control". So sometimes you'll see "policy management", sometimes you'll see "policy control". A literal translation that would make sense to people is "target and means management". (The significant aspect of hoshin planning is its strong focus on the means, the process by which targets are reached.) Another name is "Management by Policy (MBP)" which is used in Japan to distinguish it from "Management by Objectives (MBO)".

Until recently, GOAL/QPC called this system Management by Planning, because they thought at least people would get the idea that it was about planning and not about personnel policies. But people felt that Management By Planning sounded watered-down and failed to convey the idea of top management setting the fundamental direction for the organization. So GOAL/QPC now uses the name hoshin planning. Two of the major U.S. corporations that use this system, Hewlett-Packard and Procter & Gamble, call it Hoshin. Florida Power and Light, which has used a modified hoshin planning for three or four years, has called it policy deployment.

To understand hoshin planning, one must understand Total Quality Management, the management system of which hoshin planning is a part.

PART 2: Hoshin's Role in Total Quality Management

Part 2 includes the Management Revolution occurring today, a definition of Total Quality Management, and the evolution of Total Quality Management in Japan and the U.S. The Total Quality Management system is examined in terms of systems tools and organizations.

A Management Revolution

Prior to the introduction of Total Quality Management, quality efforts were based on the ideas of Frederick Taylor. Around 1900, Frederick Taylor concluded that not every supervisor was smart enough to figure out the best way to do a job, and decided that an engineer should plan the job and the supervisor should carry out the engineer's plan. In the 1920's and 1930's quality control was added to check that manufacturing was doing the job the way the engineers had set it up.

This system was a tremendous success. Bethlehem Steel has researched the early 1900's in their company and found documentation of 200 and 300 percent increases in productivity using this system. During the century, however, the education level of supervisors and production workers improved, but the system wasn't changed. Actually the system was changed-- we put barbed wire on top of the walls between departments.

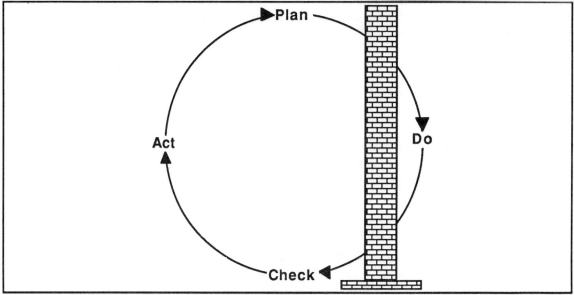

FIGURE 1.1

Today we are taking down the wall, in terms of management. The Plan-Do-Check-Act (PDCA) cycle, also known as the Deming or Shewhart Cycle, is the cycle of activity. A company Plans to do something, Does it, Checks the results, and Puts it into Action or Adjusts.(See Figure 1.1)

We now have all employees in all departments involved in the PDCA cycle, improving or maintaining their own jobs. And if quality, yield, cost, procedure or systems are being examined, it's not just Quality Control. It involves the whole realm of management.

(Note the word "maintaining". We'll come back to that. Some people have described TQC as continuous improvement, but TQC equals continuous improvement plus maintenance. These two concepts must be tied together in order for planned results to be attained.)

> # Company-Wide
> # Continuous Improvement Process
>
> | PARTICIPANTS | ACTIVITIES | RESULT |
> |---|---|---|
> | All Employees | Improve or maintain own jobs and tasks in all variations | Allow company to supply products to its customers |
> | All Departments | • Quality
• Yield
• Procedure
• Cost
• System | • Most Economic
• Useful
• Competitive
• Best Qualified |

FIGURE 1.2

Definition of Total Quality Control (TQC)

The result and definition of TQC is that companies supply products and services that are most economical, useful, competitive, and of the best quality. In the U.S., TQC is often called Total Quality Management.

Japan's Quality History

In the late 1970's, Juran drew a line graph indicating that in the 1950's the U.S. was superior to Japan in quality. Japan improved and surpassed the U.S. in the early 1970's. Juran's ideas were not rooted in any fact. They were rooted in his and Deming's perception that the U.S. was far superior in quality in the 1950's and 1960's and that Japan was deplorable, and that Japan began to improve while the U.S. stagnated, and somewhere in the early 1970's the two were equal. There's no statistical basis for this, it's simply his perception of the quality movement. In Figure 1.3 the author expands the chart and suggests that the U.S. may now be beginning to narrow the gap.

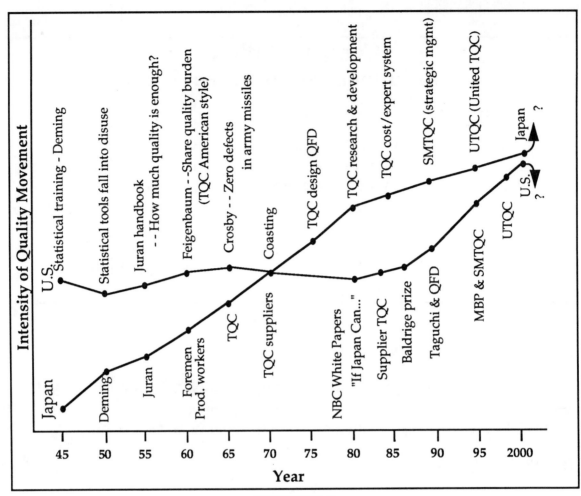

FIGURE 1.3

Some of the TQC milestones in Japan are as follows: In the 1940's Shewhart's book was studied, followed by Deming in 1950, and Juran in 1954. Deming taught engineers quality methods, Juran taught managers; foremen were taught in 1960, production workers in 1962, in 1965 all employees were included in the new concept of TQC, and suppliers were involved in 1970. TQC of Design was also called Quality Function Deployment in 1975; TQC of Research and Development began in 1980; TQC for Cost Control Related to the Yen Appreciation and TQC Expert Systems in 1985.

In a 1988 book, Masao Kogure, who is one of the leaders of the quality movement in Japan, forecasts SMTQC for the 1990's. SMTQC stands for Strategic Management Total Quality Control and is related to United Total Quality Control. Both of these play a part in the development of the optimal TQC system in a multi-national company, in which there are similarities in the quality program corporate wide, but the total quality program is customized to be the best it can be in each company. It is predicted that SMTQC and UTQC will be the major growth areas for TQC in the future.

U.S. Quality History

In the U.S. there was statistical training by Deming during WWII, but it was abandoned after the war. Juran's handbook, Feigenbaum's TQC, and Crosby's Zero Defects were the next milestones. The NBC White Paper in June of 1980 - "If Japan Can, Why Can't We?" was seen by a lot of corporate leaders of the country, and began renewed quality efforts which were quickly pushed on suppliers. In the late 1980's Taguchi and QFD came into play, followed by the Baldrige prize in 1988.

In the 1990's Strategic Management TQC and United TQC will be the growth areas. My hypothesis is that the growth in Japan is slowing down, our growth is accelerating, and in the 1990's we may catch up. But notice the upward curve in Figure 1.3. It's possible that as we make progress, the Japanese will find new breakthroughs and keep their leadership position, and it's also possible that Americans will lose interest in TQM and shift to next year's 'instant pudding'.

The TQC System

The systems that comprise TQC are extensive. If you visited the leading organizations in Japan and said, 'I want to learn about TQC, who should I talk to?', the answer would be that there is no single person you should talk to, that you need to talk to six or seven or eight different people, all of whom have a specialty, all of whom will give you a piece of TQC. Through listening to a lot of these specialists, reading their books, and translating their writing, the author has put together a schematic (TQC Wheel) that summarizes what the systems are, and the interrelations of those systems for Total Quality Control.

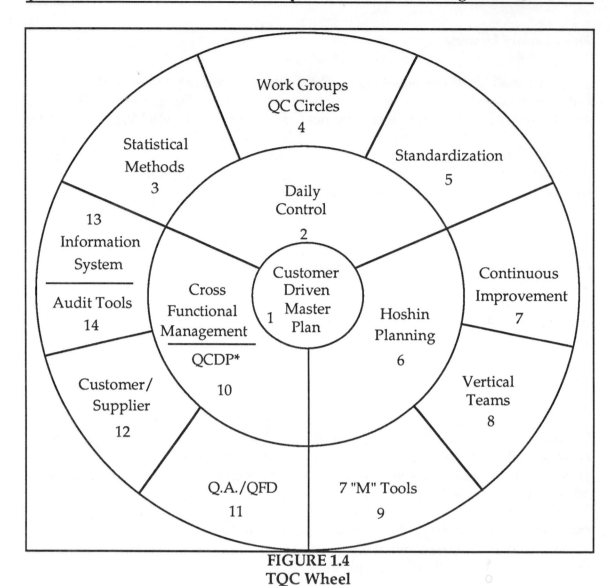

FIGURE 1.4
TQC Wheel

What is T.Q.C.
1. Customer-Driven Master Plan
2. Daily Control
3. Statistical Methods
4. Work Groups/QC Circles
5. Standardization
6. Hoshin Planning
7. Continuous Improvement
8. Vertical Teams/Customer
9. 7 "M" Tools
10. Cross-Functional Mgmt.
11. QA/QFD
12. Horizontal Teams-QCDP
13. Information Systems
14. Audit Tools

*Quality, Cost, Delivery, Profit (or Product)

TQM Vision

1. **Imagine** an organization that knows what customers will want 5-10 years from now and exactly what they will do to far exceed all expectations.

Chapter 1: Hoshin Planning and TQM   p.1-9

2. **Imagine** an organization where each employee knows what he needs to do to make the organization run smoothly. His actions are documented, audited, and updated daily as changing situations require.

3. **Imagine** an organization where each employee manages by facts and knows how to analyze problems by using simple tools to understand variability and data.

4. **Imagine** an organization where each employee generates 100-200 suggestions per year (2-4 per week), of which 95% is implemented, and joins with the work groups to maximize progress.

5. **Imagine** an organization where everybody knows the most important variables to control in order to satisfy customers, guarantee effectiveness and efficiency, and where these standards are documented and updated daily.

6. **Imagine** an organization where the president sets the two or three most important goals for the year, every manager knows these goals, and the two or three most important tasks to help achieve these goals, and each manager has measurable milestones for these activities which he personally audits monthly, documents, and sends up through the organization to enable diagnosis and improvement.

7. **Imagine** an organization where each employee understands not only how to do his job but also how to significantly improve his job on a regular basis.

8. **Imagine** an organization where all problems and challenges are met by a team of the most appropriate people, regardless of their levels or jobs within the organization.

9. **Imagine** an organization where all managers and staff people use effective and simple planning tools on a regular basis to do a better job.

10. **Imagine** an organization where cross-functional teams assure that quality, cost, efficiency, services, and profit are managed on a consistently high level throughout each business unit of the organization.

11. **Imagine** an organization where quality assurance and reliability are managed effectively on a daily basis and the total organization is thoroughly familiar with customers.

12. **Imagine** an organization where each employee knows all the people who supply him or her with data and material and gives these

people clear, concise advice on how to improve that data and material and also where each person knows all his customers and seeks ways to meet expectations in providing data and material.

13. **Imagine** an organization where all information smoothly and concisely flows daily to the people who need it.

14. **Imagine** an organization where improvement activities are audited at each level of the organization to assure that each employee reaches his or her full potential.

Customer-Driven Master Plan

At the heart of the TQC wheel is a customer-driven master plan. This includes the organization's 5-10 year plan, rooted in its customer. The plan includes how the company will transform itself over the next 5-10 years to be a market leader in its area of specialty. The master plan focuses on product, organizational effectiveness, and, in many organizations, profit. It's the strategic plan, if you will, rooted in the customer. Surrounding the master plan are three major systems: daily control, hoshin planning, and cross-functional management.

Daily Control

Daily control is the part of TQC in which each employee knows clearly and simply what he or she has to do in order for the organization to run smoothly.

Supporting daily control are tools, organizations, and sub-systems. One of the tools is statistical methods. Statistical methods help people understand and control variation. They help people operate and make decisions by facts. They also help people recognize what they are doing in their daily activities that is effective and what they are doing that is ineffective.

Organizational tools include work groups and quality circles. These provide a team approach in which people get together to work on problems and to get better at what they are doing.

Standardization is the written documentation of all procedures in a simple and clear way, updating those standards on a regular basis, even daily.

In the States we've used statistical methods and quality circles to some extent, but could do a lot more in terms of the daily control and standardization.

Chapter 1: Hoshin Planning and TQM p.1-11

Hoshin Planning

The second major category of TQC is hoshin planning. Hoshin planning helps orchestrate the direction of the company. In this system each manager selects his three or four most important activities which tie into the top three priorities of the company.

Related to hoshin planning are the tools for continuous improvement, the tools for breakthrough, thinking of ways to do things better, implementing ideas, and acquiring the habit of continuous improvement and annual breakthroughs. An organization's vertical teams, customer-supplier relationships, and the Seven Management Tools (described below) also help in this effort.

Cross-Functional Management

Cross-functional management is another facet of TQC. While hoshin planning is fundamentally a vertical activity, cross-functional management focuses on the horizontal activity in an organization. If a company wants to deliver a quality product to its customers, it must make sure that the purchasing, manufacturing, sales, and service organizations all have consistent, integrated quality efforts pertaining to scheduling, delivery, plans, and so on.

Quality, cost, delivery, product, and profit (QCDP) are the five most common areas in which cross-functional management takes place. The quality assurance system assists this by controlling quality horizontally. Quality function deployment is a subset of that; it is a tool used to find out what the customer wants and to get that information to all the right people in the organization. Using this system, an organization gets the three or four most critical needs of customers nailed down in such a way that it's better than any of its competitors, and gains market share as a result.

Horizontal Teams, Information Systems, and Audit Tools

Another part of the TQC wheel is horizontal teams. Horizontal teams are people communicating about quality, cost, delivery, productivity, etc., so that they are in sync. The information systems and audit tools make it possible to convey that information to all the right people and to audit the progress.

There are a lot of other systems in TQC, but those are some of the more important ones. They are presented in a way to show some of the inter-relationships and the mix of tools, organizations, and sub-systems. This doesn't mean that you shouldn't use statistical methods for continuous improvement, or statistical methods for quality function deployment; it's just a way of showing how the pieces can fit together.

The following section highlights some of the key aspects of the wheel.

PART 3: Key Aspects of Total Quality Management Related to Hoshin Planning

Part 3 gives additional information on some of the key aspects of Total Quality Control: the Customer Driven Master Plan, the Seven QC Tools (flow chart, pareto, etc.), and the Seven Management Tools (Affinity (K.J.), Tree, Matrix, etc.).

The Customer-Driven Master Plan

A couple of years ago, the author gathered the detailed master implementation plans of a half dozen Japanese companies who had won the Deming Prize, laid them on top of each other, and asked, 'what are their consistent elements in terms of their master implementation plan?' Figure 1.5 shows the results.

The first phase of the master implementation plan is the first year and a half training of twenty to forty percent of the top and middle managers, doing customer assessment, having project teams, and finding out what some of the problems are in the organization. This includes developing a vision of where the organization is going and having a clear understanding of who the organization is today. The idea of where the company is going is spread out over three to five years with a plan for each year, and includes the system changes needed to make the organization competitive in five to ten years. The training occurring at the same time as changes must also be examined so that the training is tied to change.

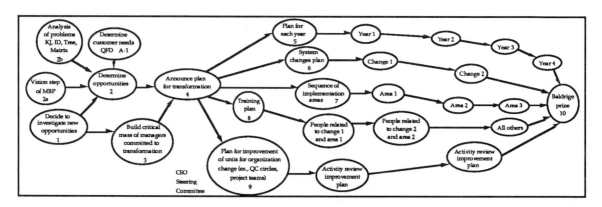

FIGURE 1.5

Chapter 1: Hoshin Planning and TQM p.1-13

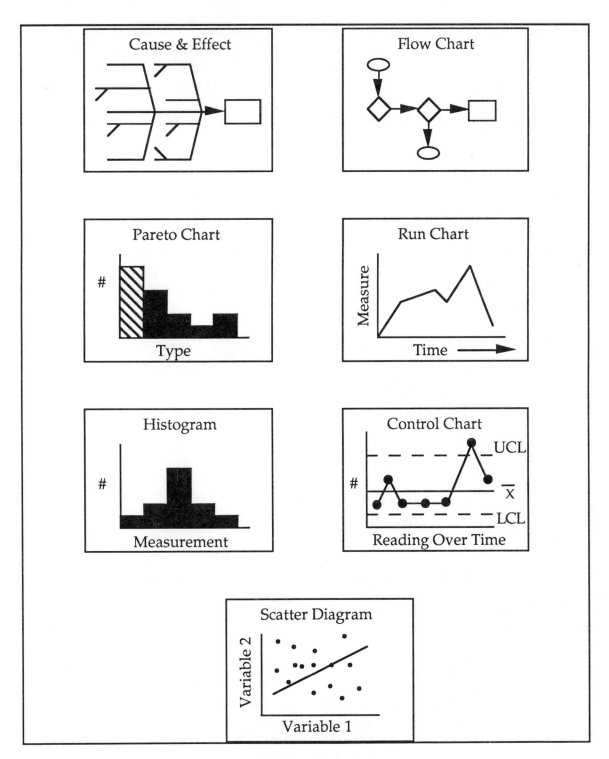

The Seven QC Tools
FIGURE 1.6

Statistical tools are needed for the plan. Some tools are: The Cause and Effect Diagram, which indicates effects and the causes and how they interrelate; the Flow Chart, which shows the way things go through the organization, the way they should go, and the difference; the Pareto Chart, which indicates the rank of causes by frequency or severity; the Run Chart, which shows whether key indicators are going up or down and whether that's good or bad; the Histogram, which shows variation in certain measured units; the Control Chart, which indicates whether the organization has system problems or special causes; and the Scatter Diagram, which shows the relationship between two variables.

Seven Management Tools

One of the major breakthroughs in hoshin planning in this decade was the development of the Seven Management Tools in 1982 by Nayatani. Nayatani put these tools together in the 1970's for hoshin planning and implementation of TQC. Two of the Seven Management Tools are for general planning, three are for intermediate planning, and two are for detailed planning.

The Affinity Chart   The Interrelationship Digraph
General Planning
FIGURE 1.7

As a planning example, think of the food industry. Universal Foods may use these tools when considering cheese. They could conduct surveys about what people look for in cheese, they could interview people, send out surveys, and read trade magazines in terms of cheese and different preparation. Then they would have a lot of information about cheese. What would they do with it?

One thing they could do is make an Affinity Chart. They would put every bit of information on 3M Post-It Notes. Then they'd group together pieces of information that are related, physically moving the Post-Its until there were groupings of cards. Next they'd select or make Header Cards to act as headings for each group. Now all this information about cheese would be organized.

Another idea is to use the Interrelationship Digraph or the Relations Diagram, as some people prefer to call it. In this case you would take the same bits of information and ask 'which one causes or influences another one?' Then you would draw an arrow to indicate influence. The items that have the most arrows going in or coming out are the critical items because they influence or are influenced by a lot of things. The items that have mostly arrows going out are also important.

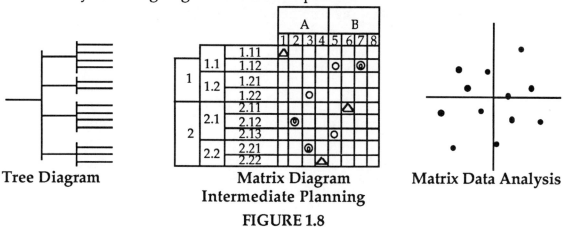

Tree Diagram **Matrix Diagram** **Matrix Data Analysis**
Intermediate Planning
FIGURE 1.8

In intermediate planning (continuing with the cheese example), consider the categories or header cards from the Affinity Chart as the major concerns in cheese, and ask, 'how can we address those?' Then detail that. This is where a Tree Diagram would be used. You might say to yourself, 'who in the organization is responsible for taking care of these items, for making sure that we do a good job in meeting customer demands?' This is when you'd see who is responsible for each of these major concerns in the cheese business and where the matrix diagram would be used.

The final chart is slightly different from the others, and in Japan is used in a very mathematical sense. It portrays data on an x and y axis so that you can see relationships. Using our cheese example, you might plot different kinds of cheese. You would consider some of the variables of cheese that would help you compare your product to other products, such as fat content. High and low fat content would be on the x axis. High and low salt content would be on the y axis. This system makes it possible to plot each of your products and your competitors' products and think about what new products you want to introduce, advertise, etc. This is called Matrix Data Analysis, otherwise known as market segmentation.

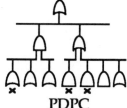 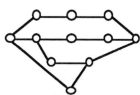

PDPC **Arrow Diagram**
(Process Decision Program Chart)
Detailed Planning
FIGURE 1.9

One example of detailed planning is the PDPC or Process Decision Program Chart, which is used if you are doing something for the first time. Let's say you want to introduce a new packaging for cheese that keeps the cheese fresher than any other packaging, and is easier to open than any other packaging. It's a new packaging for cheese. You would list all the steps for doing that, all the things that could go wrong, and the countermeasures for all the things that could go wrong. PDPC is a contingency planning tool, which would help map out the introduction of this new cheese wrapping.

The arrow diagram is another detailed planning tool which examines those things that you do over and over again. It might be the steps after you've developed the new wrapping for the cheese or how you set up the manufacturing process and actually begin producing the wrapping. Those would be things you've done many times before and know how to do. This procedure shows which things may be done simultaneously, and what the critical paths are. It's very much like the PERT (Program Evaluation and Review Technique) or CPM (Critical Path Movement) Charts.

In the States we've made PERT and CPM charts better but more complicated, so that fewer people use them. The Japanese have made these tools simpler so that more people use them. Because you get so much benefit from a small amount of energy or activity, they should be popular.

PART 4: Daily Control and Hoshin Planning

One final note concerning the TQC wheel. There is a very important relationship between daily control and hoshin planning.

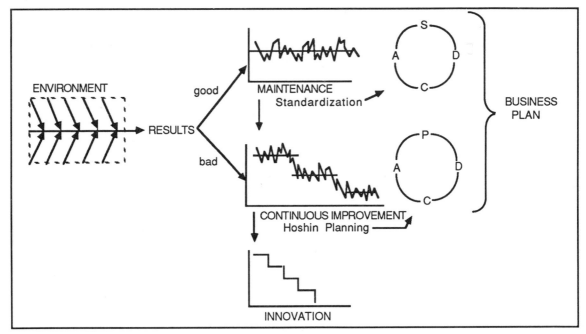

FIGURE 1.10

In an organization, activities happen in an environment consisting of competition, customers, government, and other factors. As a result of those activities in that environment, you produce certain results. Of those results some are good and some are bad. You want to maintain the results that are good, so you have an activity called maintenance. You have some results that are bad. You want to change the bad results and make improvements, so you have continuous improvement. The system for maintaining this maintenance is standardization.

Maintenance is controlled by the standardization system, and the method for doing that is the SDCA cycle, in which we Set the standard, implement or Do it, Check to see if we've followed it, and then put it into Action or Adjust. This is the way in which we continue on a daily basis to do all the right things.

Our daily audit of that is called daily control. Daily control makes sure that everyone is doing his job as he should, and that all the documentation is up to date.

Hoshin planning is the system that orchestrates continuous improvement and breakthroughs. It picks the area that needs improvement, makes sure all the right people get involved, and that the improvement is implemented.

The control is a monthly audit, and the method is the Plan-Do-Check-Act (PDCA) cycle. The business plan of the organization includes both continuous improvement and innovation.

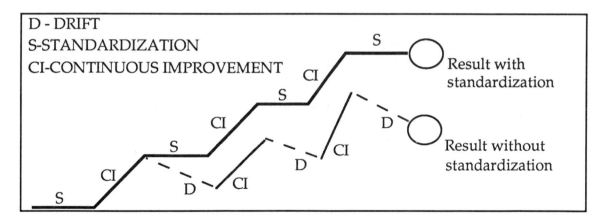

FIGURE 1.11

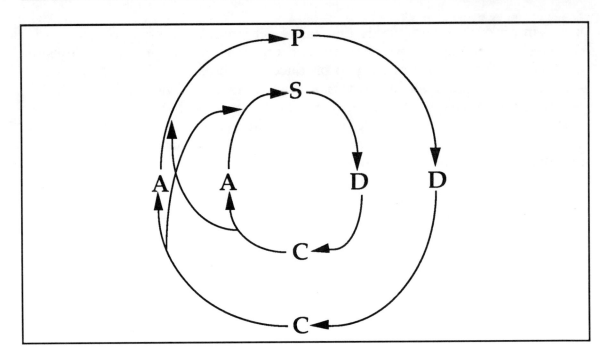

FIGURE 1.12

The Plan-Do-Check-Act cycle and the Standardize-Do-Check-Act cycles are related. They are the daily activities the company has if it is operating according to a standard. If the results are good, then you maintain that standard in the SDCA cycle. If the standard is not giving you desired results, you need to improve it and focus on the improvement circle of the PDCA. If you're in the PDCA circle, you're planning innovation or improvement. If you try it and are successful, then you want to standardize that, to make sure that you keep doing that.

In some organizations, if there is a crisis today, you work on that crisis. You make an improvement, but tomorrow has another crisis, so you shift your effort to that new crisis. Meanwhile the thing that you fixed today goes out of whack because people aren't paying attention to that, they're paying attention to something else. You get caught in a fire-fighting mode. You keep having all kinds of frenzied activity, but after two or three years you look at where you are, and you're not much different from where you were two or three years ago. You must have a blend of standardization and continuous improvement for significant and lasting progress.

Chapter 2
Planning Problems and Hoshin Solutions

This chapter will look at the problems U.S. organizations sometimes encounter in planning. It will then review the principles of hoshin planning and the basic system hoshin planning uses to orchestrate continuous improvement.

PART 1: Some Problems with Planning

U.S. students of hoshin planning have observed the following problems with planning:

- No Plan

The first basic problem is having no plan. In some organizations activities just evolve based on a propose-dispose activity. Subordinates propose activities and the boss votes it up or down. One thing is clear. No matter what you do or fail to do, it will get you there if you don't know where "there" is.

- Short Range Focus

Many companies ask customers to place orders early so that the end of the month or quarter or year will look good. This means people work overtime at the end of the month to get extra orders out and sit around or get laid off at the beginning of the month when there is nothing to do. Much of this short term thinking is generated by the stock market and the felt importance of meeting stock analysis expectations. It is of course highly inefficient, wasteful, and detrimental to the long range interest of the company.

- Difficult to Measure Success

Many goals are not stated in a manner in which they can be easily measured. If goals are not measurable it is quite difficult to tell whether they have been achieved or not.

- Don't Measure

Sometimes goals are measurable but they are not measured. Sometimes the goals are filed away and forgotten. The next year, armed with a blank sheet of paper, people develop a new plan without any reference to last year's performance.

- <u>Language Problems</u>
Sometimes the goals are not written clearly. They may be too lengthy, too complicated or the words may be ambiguous. It is possible for people to understand the words in different ways, especially if they did not participate in developing the goals.

- <u>Gets Filed</u>
If the plan is filed, it may soon be forgotten. It is certainly not reviewed or updated. It may be dragged out for review at the end of the year, for developing a new plan, or it may just be "permanently filed".

- <u>Fragmented</u>
In larger organizations there may be plans for different business units that are not connected in any way. This can also happen from department to department or from person to person.

- <u>Long Range by Corporate Only</u>
In some cases the long range plan is devised by the corporate group. A consultant may come in to help devise the plan. Sometimes the plan is kept confidential. It is difficult for employees to contribute to the plan if they do not know what it is.

- <u>Long Range Not Related to Daily</u>
Planning is sometimes seen as a different activity from daily work. People separate to do planning and then come back together to do their regular work. There is no tie-in with daily activities. Consider an alternative in which everyone in the organization knows the five year plan, the three or four key things that they have to do to move that plan along, and each of them ties it into their daily activities.

- <u>Can't Handle Emergencies</u>
Sometimes things change inside or outside the company that effect the plan. It would be useful if each person in the organization were watching for these changes and knew what to do if they were to happen.

- <u>New Manager</u>
A manager may be replaced during the year. There may not be sufficient documentation on how the plan has been implemented to date. This makes it difficult for the new manager to pick up the plan and may significantly effect the progress.

Chapter 2: Planning Problems and Hoshin Solutions   p. 2-3

- • Don't Plan for Resources
People may not allocate money and people consistently with the plan. This can lead to a significant amount of frustration.

- • People Plan for Others
Sometimes planning is a staff function in which plans are developed for other people. Outsiders may not know enough about the situation to develop a good plan.

- • Unrealistic Plan
A stretch goal is sometimes far beyond anyone's reach but is used with the hope that more progress will be made than would have been achieved otherwise.

- • Plan as a Weapon Not a Tool
Sometimes a plan is a veiled or not so veiled threat.

- • The Plan is Poorly Communicated
In some cases the plan is seen as company confidential and cannot be communicated to the people who are in a position to take action on it.

PART 2: The Principles of Hoshin

Imagine an organization that knows what customers will want five to ten years from now and exactly what they will do to meet and exceed all expectations. Imagine a planning system that has integrated PDCA language and activity based on clear, long-term thinking, a realistic measurement system with a focus on process and results, identification of what's important, alignment of groups, decisions by people who have the necessary infor-information, planning integrated with daily activity, good vertical communication, cross-functional communication, and everyone planning for himself or herself, and the buy-in that results. That is hoshin planning.

Hoshin Planning and MBO

How is hoshin planning different from MBO?

	MBP	MBO
Purpose	• Kaizen - Continuous <u>Improvement</u> • Organization team-oriented	• Management <u>control</u> • Personnel oriented
Organization	• All employees • Bottom up/top down blend	• All related people work together • Trouble shooting decision by objectives
Objective	• Quality First	• Profits, cost first
Methodology	• Flexibility • Every member participates	• Politics • Either top-down or participative
Key procedures	• Results through process • Organizational self-diagnosis • No direct link with personnel	• Results through targets • Linked to job evaluation and salary

Adapted from Kozo Koura, JUSE consultant, ICQC Proceedings 1987[1]

FIGURE 2.1

Although the Japanese often compare hoshin planning and MBO, it is difficult to make such a comparison in the States because of the different kinds of MBO implementation. There are people, particularly in smaller organizations, who have enough contact and enough people doing several different jobs, that have had success with MBO. In a situation like that, if you tell people that you have something to replace MBO, people will say, "I don't want to replace it, I like it, it's good. Don't take away what's good". So this is just a caution. If you start out with what's wrong with planning, and criteria for good planning, and then lead into it, the author has found you set a more receptive atmosphere.

Principle 1: Participation by all Managers

One of the principles of hoshin planning is participation by all

---

[1]Note: MBO (Management by Objectives) is described here as it is implemented in Japan. There are over 12 different models in the U.S., so the reader may be familiar with a different approach.

Chapter 2: Planning Problems and Hoshin Solutions

managers in setting a five year vision. There are very few organizations in which every manager is asked to give his or her input in terms of a five year vision. America has more of a problem with this than the Japanese because the organizational structure of the Japanese is simpler. They have typically, even in huge organizations of hundreds of thousands of people, four levels: top management, plant management, department management, and section management, with some modifications. But in the U.S. we have organizations consisting of sometimes 10, 15, 20 different levels.

In hoshin planning, the top group - the president and his or her staff - put together a vision, and send that down to the organization for comment. In Japan they ask, "If this were the vision, what would you do?" In the States people may want to be more direct and state what's right or wrong with the vision, and then send that information back up, so that there is a basis for deciding whether or not people understand the vision, or whether it will produce the desired results.

Principle 2: Individual Initiative and Responsibility

Each manager sets personal goals or objectives. (For some people it's a goal, for others it's an objective. There is a Glossary of Terms in the back of this text that may help define key words and phrases, but no glossary will be able to resolve some words whose usage has been muddied over the years.)

In hoshin planning, each manager sets monthly and yearly targets for himself. (Note that we're just talking about management, we're not talking about production workers. In the Japanese TQC model, quality circles are used to combine management with production workers.)

In an autocratic organization the boss sets the target. In a participative organization, the boss and subordinates sit down together and come to some agreement as to what the targets will be. In hoshin planning, the subordinate sits down by himself or herself and decides what he or she thinks the target ought to be, and then gets together with others in the organization to align those targets. This system focuses on the individual developing his own ideas of what needs to be done, and then integrates them. Goals are measured by the individual manager who sets them for himself or herself.

Principle 3: Focus on Root Causes

Each manager sets monthly targets, and each month evaluates the progress with a focus on getting past symptoms to root causes. The five why's, the cause and effect diagram, and ongoing analysis are some of the tools that make this possible.

Principle 4: No Tie to Performance Appraisals

In Japan, when implementing hoshin planning, there is no tie to performance evaluations or other personnel measures. The companies that have implemented this in the States, such as Hewlett-Packard and Florida Power and Light, are currently not following this principle. They have tied it to performance. HP says this is part of what people do, so it should be measured. In Japan, companies want everyone to "buy in" so that it will be a team effort. They feel if it is part of performance appraisals, there may be problems. So the Japanese keep that separate.

Dr. W. Edwards Deming has suggested dropping performance appraisals. In all too many companies, performance appraisals are a lottery with results based on chance as well as effort with no way to distinguish between the two. The author's opinion is that in transforming to a new management system, performance appraisals will be the last thing to go (successful pilots in the Department of Defense and at Ford Motor Company notwithstanding).

Dropping performance appraisals without having a change in culture, or different systems in place that will support a high level of performance in a different way, will cause a problem. Deming first warned of the negative effects of performance appraisals in 1984 as a result of work that he did in a couple of large U.S. corporations. The author has also worked with one of those corporations and learned that in this organization, which is one of the largest ten corporations in this country, that if you help another employee, there is a direct correlation to a drop in your pay. It may sound crazy, but that is indeed happening.

Most organizations have figured out how to deal with people on a regular basis, and some of that is productive. It's better that you don't try to tear down the system until you have something to replace it with. In my opinion, hoshin planning is the something to replace it with.

Principle 5: Quality First

Focus on quality first not profit first, focus on the planning process system and not the target. Good results become a regular by-product if this principle is followed.

Principle 6: Catch Ball

The catch ball principle is a communication idea in which you communicate extensively vertically and horizontally. The thing to remember about a game of catch is that the ball not only goes back and forth, it goes back and forth often. Catch ball assures communication and understanding.

Principle 7: Focus on Process

Each manager sets a numerical goal for what he's going to accomplish for the year, and each month during the year he evaluates his progress and what helped or hindered it. The focus of hoshin planning is on the process and assessment of why what you did was helpful or unhelpful in moving you toward the target.

Figure 2.2 displays common planning problems and the hoshin principles that may be used to rectify such situations.

Problem	1 Participation by all managers	2 Individual initiative and responsibility	3 Focus on root causes	4 No tie to performance appraisals	5 Quality First	6 Catch ball understanding	7 Focus on process
No plan	√	√			√		
Short-term focus				√	√		√
Difficult to measure success		√					
Don't measure		√			√		√
Language problems	√	√				√	
Gets filed		√				√	√
Fragmented	√					√	
Long range by corporate only	√	√				√	
Long range not related to daily		√	√			√	√
Can't handle emergencies	√	√		√	√	√	
New manager						√	√
Don't plan for progress	√	√	√		√		√
People plan for others	√	√				√	
Unrealistic plan		√	√	√		√	√
Plan as a weapon, not a tool	√	√	√	√	√	√	√
Plan poorly communicated	√	√				√	

FIGURE 2.2

PART 3: The Hoshin Planning System

This section describes the hoshin planning system.

The hoshin planning system involves continuous improvement of planning. In its simplest form it involves a plan, execution, and audit. In its more detailed form it includes a long range plan (5-10 year vision), a one year plan, deployment to departments, execution, monthly audits, and the president's annual audit.

FIGURE 2.3

1. *The five year vision* includes a draft plan by the president and executive group, an improvement plan based on internal and external obstacles (environment), and revision based on input from all managers on the draft plan. This enables top management to develop a revised vision which they know is understood by all and which will produce desired action.

2. *The one year plan* involves the selection of activities based on feasibility and likelihood of achieving desired results. Ideas are generated from the five year vision, the environment (including customer demands as determined by Quality Function Deployment), and ideas based on last year's performance. Techniques such as storyboarding, surveys for data collection, pareto analysis, and market trends may also be helpful.

Chapter 2: Planning Problems and Hoshin Solutions   p. 2-9

Ideas based on the five year plan analyze obstacles to the five year plan and develop a systematic draft plan.

The review of last year looks at evidences of management inadequacy, compares planned and actual results, and develops a systematic plan of what should be done.

The tentative plans developed by looking at vision, environment, and last year are critiqued. A number of different criteria may be used, but some of the more popular are: predicted effect, return on net asset, service to trade, simplicity, total system effect, and engineering and resource feasibility.

Each potential plan is rated against the selected criteria using a number or letter classification and a decision is made on the best action plans.

Some of the concerns people have with this approach to a one year plan are how to handle conflicts between the three sets of goals (develop a synthesis), how to decide who will evaluate it, and the time consuming aspect of the process. On the positive side, people praise its systematic approach and thoroughness.

3. *Deployment to departments* includes the selection of optimal targets and means (See Figure 2.4). It focuses on the identification of key implementation items and a consideration of how they can systematically accomplish the plan. The individual plans so developed are evaluated using the criteria that were also used in Step 2 above (e.g., feasibility, likelihood to produce desired effect).

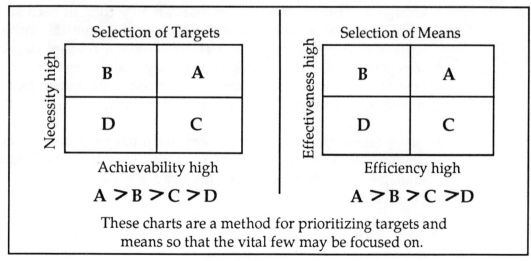

FIGURE 2.4

The appropriate measures for the selected plan are developed using the matrix of measures and means. Finally, a flag system chart is developed which identifies the current state, the plan, and the scope of the line of improvement expected during the twelve months of the one year plan. The final plans are submitted to a committee or facilitator whose job is to help iron out vertical and horizontal inconsistencies in the various plans.

Some people feel that this approach may be difficult to integrate with existing systems or that the number of potential measures may be unmanageable and plans may be limited only to what is measurable. Also, some worry that in an organization accustomed to top-down control, the measures developed by managers to monitor subordinates' improvement efforts may be turned on them to control their actions in what might become a repressive management style.

On the positive side, people believe that hoshin planning is superior to existing deployment because it is bottom-up and enables one to more easily measure the right things; the measures are visible line charts rather than columns of numbers. It builds a basis for evaluation as well as quick and regular reviews.

4. *Execution* is the implementation of the detailed PDPC and arrow diagram (See Fig. 1.9) or PERT chart developed to accomplish the plan. It focuses on the power of contingency planning. The steps to accomplish the task are identified and arranged in order. The things that can go wrong at each step are listed. Countermeasures are listed for each potential problem. Appropriate countermeasures are selected. A revised step by step plan is drawn. All the contingency plans may not be possible because of resources or other constraints, but their consideration may still pay off at a later date.

Some drawbacks to this approach to execution is that it implies a lack of variation and may lead to a process being over-managed. The level of detail may cause time constraint problems.

However, hoshin planning uses simple tools for contingency planning. These tools make it possible for a large number of people to do contingency planning, thus leading to clear monthly targets and increased likelihood of success. Other advantages are self-diagnosis, self-correction, and visual presentation of action. This very thorough approach reduces the likelihood of things falling through the cracks.

5. *Monthly diagnosis* is the analysis of things that helped or hindered progress and the activities to benefit from this learning. It focuses attention on the process rather than the target and the root cause rather than the symptoms. Management problems are identified and corrective actions are systematically developed and implemented.

Some people are concerned that the audit results may not be directly related enough to the business plan. The results may not directly tie into daily control and the standards that are so critical for maintenance items.

Some of the advantages of the audit process are that it is data driven, objective not subjective, and it focuses on the process.

6. *President's annual diagnosis* is the review of progress to develop activities which will continue to help each manager function at his or her full potential. The president's audit focuses on numerical targets, but the major focus is on the processes that underlie the results. The job of the president is to make sure that management in each sector of the organization is capable. The annual audit provides that information in summary and in detail.

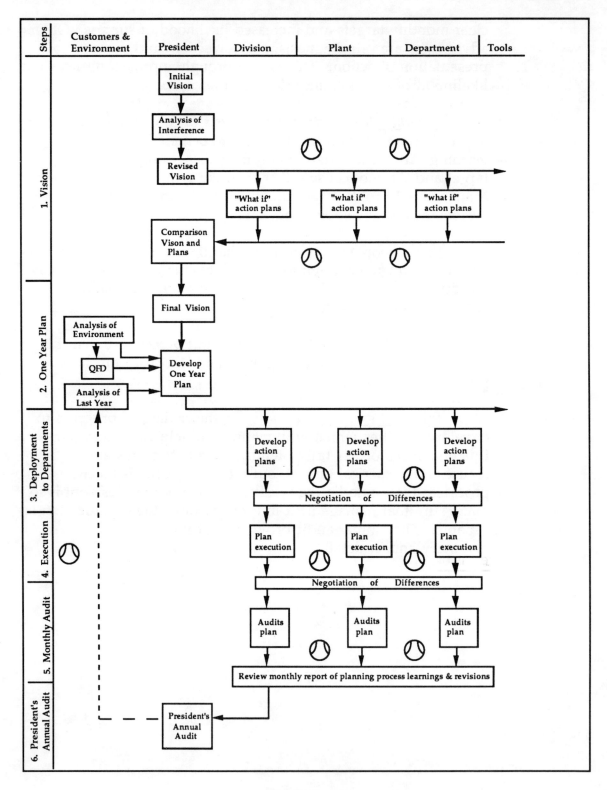

FIGURE 2.5
This people flow chart shows the interaction of activities at various levels of the organization.

How to get started:

The concept of visioning is growing in popularity. One might believe it possible to go away for two to five days and come back with a clear vision of where you are going. Actually, developing a good vision usually takes a year and a half. It includes pain, just like peeling an onion. Each continuous improvement project yields some insights into the strengths and weaknesses of the organization. As you learn from these you continually experience the pain (tears) of every additional layer of the onion until you get to the core. When you get to the core, then you have a clear idea of where you are going.

Maslow introduced the concept of a hierarchy of needs. Individuals must have their basic needs met before they can move on to higher needs.

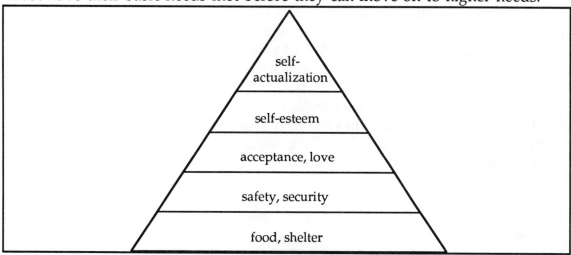

FIGURE 2.6

Organizations also have a hierarchy of needs. One way to portray these needs are in terms of the phases of hoshin planning.

FIGURE 2.7

To shift from phase 0 to phase 1, an organization must use the Seven QC Tools for process management (See Chapter 8). To shift from phase 1 to phase 2, managers must diagnose individual and organizational handicaps and take action (See Chapter 9). To shift from phase 2 to 3, managers must align individual priorities with other managers; each manager selects the top three items that he or she will personally accomplish as part of the one year plan (See Chapter 10). To go to phase 4, organizational diagnosis identifies the most important priorities for the next five years and fully integrates the organization (See Chapter 11).

About this Text

Learning to do hoshin planning is like learning to fly an airplane. It is only undertaken with an understanding of theory, numerous procedures, and lots of experience with the guide of an expert. Figure 2.8 shows a schematic of hoshin tools, systems, and phases.

This text, following that model, has begun with hoshin principles. Section two (Chapters 3-7) will deal with tools, procedures, and methods. These include the Seven QC Tools, the target-measures matrix, the flag system and other alignment tools, and the Seven Management Tools.

Section three (Chapters 8-11) describes the various levels of hoshin planning starting with a quality culture, beginning hoshin, intermediate hoshin, and advanced hoshin.

Section four (Chapters 12-13) offers implementation guidelines and case studies from both the U.S. and Japan.

Summary

It may sound complicated, but the payoff is high. Hoshin planning is recognized as having many advantages over traditional planning. It is thorough, data driven, and participative. Individuals make plans that are tied into a company vision, diagnose the processes, and compare actual and target results. The tools used are quite simple. They are easy to teach and use.

Chapter 2: Planning Problems and Hoshin Solutions

Steps	1	2	3 Deployment		4	5	6	Added Dimension
	Vision	1 Yr Plan	individual	align	Execution (Process Mgmt)	Monthly Diagnosis	Annual Diagnosis	Developmental Learnings
Phase 1						●		Mgmt by facts
Phase 2			●		●	●		Self-diagnosis
Phase 3		●	●	●	●	●		Align -3 goals each
Phase 4	●	●	●	●	●	●	●	Understand new direction
Tools:	Aim	Plan	Do	Do	Do	Check/Act	Check/Act	
QC:								
Fishbone	●		●			●	●	
Pareto	●				●	●	●	
Line	●				●	●	●	
Flow	●		●			●	●	
Check Sheet	●					●		
Histogram	●					●		
Control	●				●	●		
7M:								
KJ	●	●						
ID	●	●					●	
Tree	●	●	●					
Matrix	●	●					●	
MDA								
PDPC					●			
Arrow					●			
Alignment Tools:								
Flag		●			●	●	●	
Target/Means					●			
Cascading					●			
T/M Tree					●			
QFD		●			●			

FIGURE 2.8

Chapter 3
Problem Solving Tools

In the early 1950's, Kaoru Ishikawa of Japan began studying statistics. He noticed several likely benefits of using statistical methods. One was that statistics would contribute to the quality of Japanese goods, which at the time were generally of poor quality. Another benefit was that statistics would help engineers and managers.

It has been the author's experience that most U.S. organizations that have taken a renewed interest in statistics in the 1980's have not yet fully benefited from the use of statistics. This is because their use of statistics has not been as complete and in-depth as it should be. But it is precisely that in-depth use that makes statistics a critical part of hoshin planning. Shigeru Mizuno, one of the leading Japanese consultants in Total Quality Control, cautions clients not to make statistical tools an end in themselves, but to focus on managing for good quality.

The purpose of using problem solving tools for management is to find the root causes of things that keep people from doing a good job. Managers must work on improving the system, but they must know what is right and wrong with the system before they start tinkering with it.

A control chart will show whether the system is in or out of control and whether problems are a result of special causes outside the control limits or common causes inside the control limits. (See Figure 3.1)

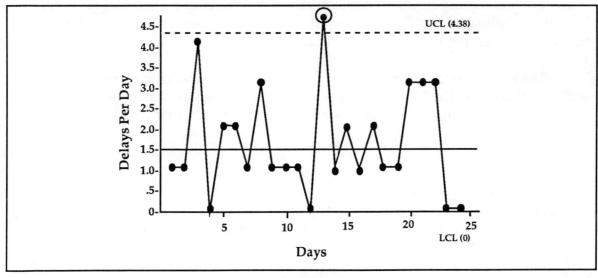

FIGURE 3.1

The Tools

The problem solving tools used most often are the following: check sheet, line chart, pareto chart, flow chart, histogram, control chart, and scatter diagram. In Japan, these are called the Seven QC Tools. They are described in Ishikawa's Guide to Quality Control, by Grant and Leavenworth in Statistical Quality Control, and in GOAL/QPC's Memory Jogger. More sophisticated tools, such as design of experiments, have been popularized by Taguchi in Japan and by George Box and William and Stewart Hunter in the U.S.[1][2]

Organizational Structures which use the Tools

Individuals Individuals can use these problem solving tools to generate suggestions. In Japan, it is not unusual to have each employee generate over 100 suggestions per year. Toyota regularly reports over two million suggestions per year, of which 95% is implemented. Part of the reason for the high implementation is that employees are allowed to try suggestions before submitting them. This results in suggestions that have already been shown to work. Also, some companies plan on supervisors spending one day a week only responding to suggestions.[3]

Groups Japanese companies usually have both work group teams and cross-functional project teams working on problems. The work group teams work on problems in their work group. The cross-functional teams work on problems that cut across organizational lines. These groups typically work on issues that relate to quality (Q), cost (C), delivery (D), or quantity and product or profit (P). The teams are sometimes called QCDP teams to identify their focus.

Problem Solving Methodology

Each company tends to develop its own problem solving methodology, or favorite sequence of problem solving tools. Table 3.1 shows that there are a number of tools that can be used for each step.

Some people think of this as an exercise of the scientific method. Others will recognize it as the Plan-Do-Check-Act cycle. By whatever name, this is a process of continuous improvement based on management by facts.

---

[1] Genichi Taguchi. *Introduction to Quality Engineering, Designing Quality into Products and Processes.* White Plains, New York, Kraus International Publications. (1986)

[2] G.E.P. Box, W.G. Hunter and J.S. Hunter (1978), *Statistics for Experimenters.* New York: John Wiley & Sons.

[3] USA Today February 3, 1989

Chapter 3: Problem Solving Tools p. 3-3

The steps are as follows:

Step	Tools	
1. To decide which problem will be addressed first (or next)	• Flow Chart • Check Sheet • Pareto Chart	• Brainstorming • Nominal Group Technique
2. To arrive at a statement that describes the problem in terms of what it is specifically, where it occurs, when it happens, and its extent	• Check Sheet • Pareto Chart • Run Chart	• Histogram • Pie Chart • Stratification
3. To develop a complete picture of all the possible causes of the problem	• Check Sheet • Cause & Effect Diagram • Brainstorming	
4. To agree on the basic cause(s) of the problem	• Check Sheet • Pareto Chart • Scatter Diagram	• Brainstorming • Nominal Group Technique
5. To develop an effective and implementable solution and action plan	• Brainstorming • Force Field Analysis • Management Presentation	• Pie Chart • Add'l Bar Graphs
6. To implement the solution and establish needed monitoring procedures and charts	• Pareto Chart • Histogram • Control Chart	• Process Capability • Stratification

TABLE 3.1

An example of this problem solving methodology may be applied to the current U.S. crisis in education.

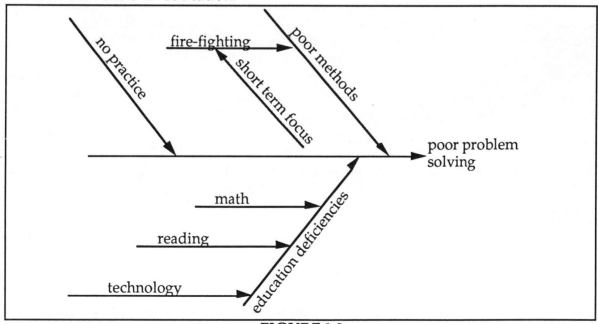

FIGURE 3.2

Let's suggest for the moment that low education level is the issue. We could examine the education level of employees.

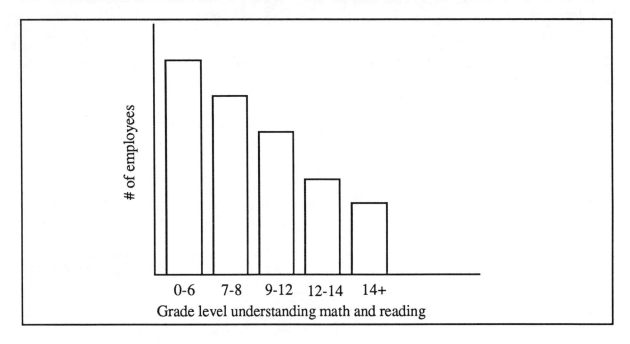

Pareto Chart
FIGURE 3.3

Motorola did such a study and is spending $50 million a year to teach seventh grade math and English to 12,500 factory workers--half its hourly employees. Kodak is teaching 2,500 people how to read and write.[1]

As George Bush began his term as president, corporate leaders were asking him to commit to improving education. But, one may ask, what is wrong with education? Is it failing to teach the basic abilities of reading and writing? Japanese education is, for example, fundamentally different. It is based on solving puzzles from kindergarten. As you solve the puzzle, you learn. Japanese quality education manuals tend to have a lot of charts and little text and the text tends not to explain the chart. As the student studies the chart, he or she learns. Japanese education is geared to solving puzzles or problems.

These are some of the issues if you consider education the top priority. But it is also possible that other causes are key. Perhaps the problem is that our management system doesn't work well and needs to be replaced to establish a long-term vision and determine priority problems. Or the problem may be a lack of standardization of procedures.

---

[1] For a detailed treatment of Japanese suggestion systems, see the *Idea Book* Productivity Press: Cambridge, MA 1988.

These problem solving tools are tools for thinking about problems, managing by facts, and documenting hunches. They are the tools of scientific management. Understanding them is the basic building block for the more advanced tools which will be covered in the following chapters.

Chapter 4
The Seven Management Tools

General Planning

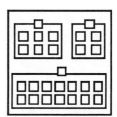

The **Affinity Diagram (KJ)** gathers large amounts of data and organizes it into groupings based on the natural relationship between each item.

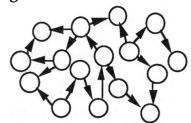

The **Interrelationship Diagraph** explores and displays interrelated factors involved in complex problems. It shows the relationships between factors.

Intermediate Planning

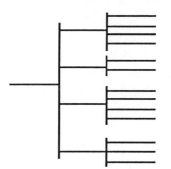

The **Tree Diagram** systematically maps out the full range of tasks/methods needed to achieve a goal.

The **Matrix Diagram** displays the relationship between necessary tasks and people or other tasks, often to show responsibility for tasks.

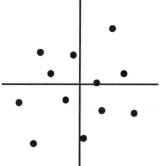

Matrix Data Analysis shows the strength of the relationship between variables which have been statistically determined.

Detailed Planning

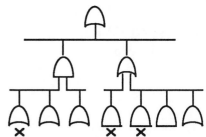

The **PDPC (Process Decision Program Chart)** maps out every conceivable event that may occur when moving from a problem statement to the possible solutions.

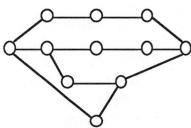

The **Arrow Diagram** is used to plan the most appropriate schedule for any task and to control it effectively during its progress.

Affinity Diagram/KJ Method

Definition

This tool gathers large amounts of language data (ideas, opinions, issues, etc.) and organizes them into groupings based on the natural relationship between each item. It is largely a creative rather than a logical process.

The biggest obstacle to planning for improvement is past success or failure. It is assumed that what worked or failed in the past will continue to do so in the future. We therefore perpetuate patterns of thinking that may or may not be appropriate. Continuous improvement requires that new logical patterns be explored at all times.

The KJ Method is an excellent way to get a group of people to react from the creative "gut level" rather than from the intellectual, logical level. It also efficiently organizes these creative new thought patterns for further elaboration. Teams may produce and organize more than 100 ideas or issues in 30-45 minutes. Think of how long that task would take using a traditional discussion process. It is not only efficient, however. It also encourages *true* participation because every person's ideas find their way into the process. This differs from many discussions in which ideas are lost in the shuffle and are therefore never considered.

When to Use the Affinity Diagram/KJ Method

We have yet to find an issue for which KJ has not proven helpful. However, there are applications that are more natural than others. The "cleanest" use of KJ is in situations in which:

- a. **Facts or thoughts are in chaos.** When issues seem too large or complex to grasp, try KJ to "map the geography" of the issue.
- b. **Breakthrough in traditional concepts is needed.** When the only solutions are old solutions, try KJ to expand the team's thinking.
- c. **Support for a solution is essential for successful implementation.**

KJ is **not** suggested for use when a problem: 1) is simple, or 2) requires a very quick solution.

Construction of an Affinity Diagram/KJ Method

The most effective group to assemble to do a KJ is one that has the knowledge needed to uncover the various dimensions of the issue. It also seems to work most smoothly when the team is accustomed to working together. This enables team members to speak in a type of shorthand because of their common experiences. There should be a maximum of six to eight members on the team.

The following are the most commonly used construction steps:

1. Phrase the issue to be considered. It works best when vaguely stated. For example, "What are the issues surrounding top management's support for a CWQC process?" There should be no more explanation than that since more details may prejudice the responses in the "old direction".

2. The responses can be recorded in one of two ways:

    a. Recorded on a flip chart pad and then transcribed onto small cards (e.g., 1" X 3"), one idea per card.

    b. Recorded directly onto individual cards by a recorder or by the contributor him- or herself. **Note:** It must be stressed that ideas should be concise and recorded exactly as stated. The aim should be to capture the essence of the thought.

3. The team should take the cards, mix them, and spread them out randomly on a large table.

4. The cards can be grouped by the team or assigned to an individual in the following way:

    a. Look for two cards that seem to be related in some way. Place those to one side. Look for other related cards.

    b. Repeat this process until you have all possible cards placed in no more than ten groupings. Do not force-fit single cards into groupings where they don't belong. These single cards ("loners") may form their own grouping or may never find a "home".

Note 1: Do not refer to these as "categories". They are simply groupings of ideas that hang together. There is a difference between the two words.

Note 2: It seems to be most effective to have everyone move the cards at will without talking. This forces team members not to get trapped in semantic battles.

    c.   Look for a card in each grouping that captures the meaning of that group. This card is placed at the top of the grouping. If there is not such a card in the grouping, then one must be written. This card should be simply and concisely written. Gather each grouping with the header card on the top.

5. Transfer the information from cards onto paper with lines around each grouping. Related clusters should be placed near each other with connecting lines. (See Figure 4.1) This is the first round of the KJ process and should be presented for additions, deletions, and modifications.

Chapter 4: The Seven Management Tools

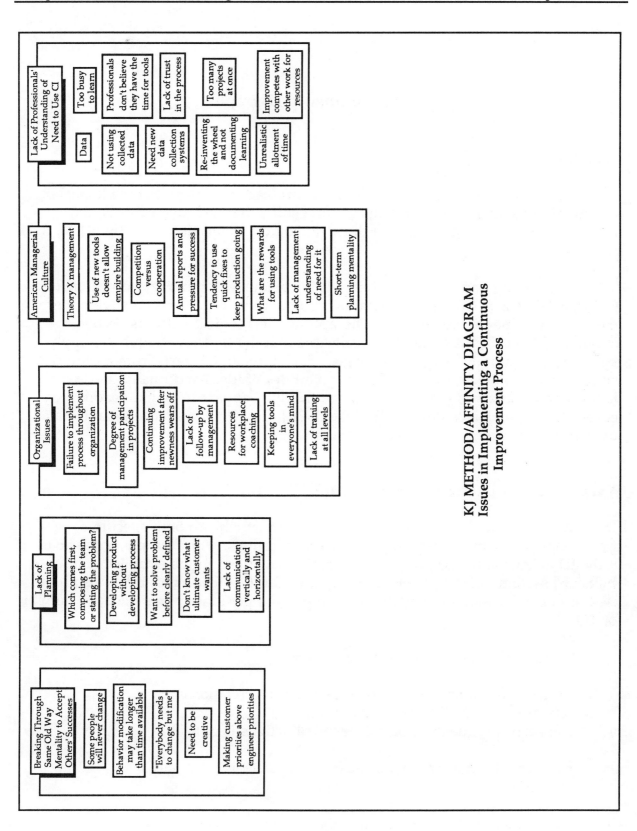

FIGURE 4.1

Interrelationship Digraph

Definition

This tool takes a central idea, issue, or problem and maps out the logical or sequential links among related items. While still a very creative process, the Interrelationship Digraph begins to draw the logical connections that the KJ Method surfaces.

In planning and problem solving, it is obviously not enough to just create an explosion of ideas. The KJ Method allows some initial organized creative patterns to emerge, but the Interrelationship Digraph (ID) lets **logical** patterns become apparent. This is based on the same principle that the Japanese frequently apply regarding the natural emergence of ideas. Therefore, an ID starts from a central concept, leads to the generation of large quantities of ideas and finally the delineation of observed patterns. To some this may appear to be like reading tea leaves, but it works incredibly well. Like the KJ, the ID allows those unanticipated ideas and connections to rise to the surface.

When to Use the Interrelationship Digraph

We have found the ID to be exceptionally adaptable to both specific operational issues and general organizational questions. For example, a classic use of the ID at Toyota focused on all of the factors involved in the establishment of a "billboard system" as part of the JIT program. On the other hand, it has also been used to deal with issues underlying the problem of getting top management support for TQC.

In summary, the ID should be used when:

a) An issue is sufficiently complex that the interrelationship between ideas is difficult to determine.
b) The correct sequencing of management actions is critical.

c) There is a feeling that the problem under discussion is only a symptom.

d) There is ample time to complete the required reiterative process.

Construction of an Interrelationship Digraph

As in the KJ diagram and the remainder of the tools, the aim is to have **the right people, with the right tools working on the right problems.** This means that the first step is to define the necessary blend of people for a group of 6 - 8 individuals.

The construction steps are as follows:

1. **Clearly** make one statement that states the key issue under discussion.
   Note: The source of this issue can vary. It may come from a problem that presents itself clearly. In this case, the ID would be the first step in the cycle rather than the KJ. The KJ is frequently used to generate the key issues to be explored in the ID.

2. Record the issue/problem statement. It can be recorded by:
   a) Placing it on the same type of card as is used in the KJ.
   b) Writing it on a flip chart.

3. To start the process, place the statement in one of two patterns:
   a) A centralized pattern in which the statement is placed in the middle of the table or flip chart paper with related ideas clustered around it.
   b) A unidirectional pattern in which the statement is placed to the extreme right or left of the table or flip chart paper with related ideas posted on one side of it.

4. Generate the related issues/problems in the following ways:
   a) Take the cards from a grouping under KJ and lay them out with the one that is most closely related to the problem statement placed next to it. Then lay out the rest of the cards in sequential or causal order.

   b) Do wide-open brainstorming, place the ideas on cards and cluster them around the Central Statement as in "a".

   c) Do wide-open brainstorming but directly onto the flip chart instead of cards. Proceed as in "a" or "b".

Note 1: The advantage of using cards is that they can be moved as the discussion progresses. The flip chart method is quicker but can become very messy if changes occur.

Note 2: When using the flip chart method, designate all the related ideas by placing them in a single lined box.

5. Once all of the related idea statements are placed relative to the central problem statement, fill in the causal arrows that indicate what leads to what. Look for possible relationships between each issue and every other issue.
   Note: At this step you would look for patterns of arrows to determine what the key factors/causes are. For example, if one factor had seven arrows coming from it to other issues, while all others had three or less, then that would be a key factor. It would be designated by a double hatched box.

6. Copy the ID legibly and circulate identified key factors to group members.

7. As in the KJ, you may draw lines around groupings of related issues.

8. Prepare to use the identified key factors as the basis for the next tool, **The Tree Diagram.**

Chapter 4: The Seven Management Tools

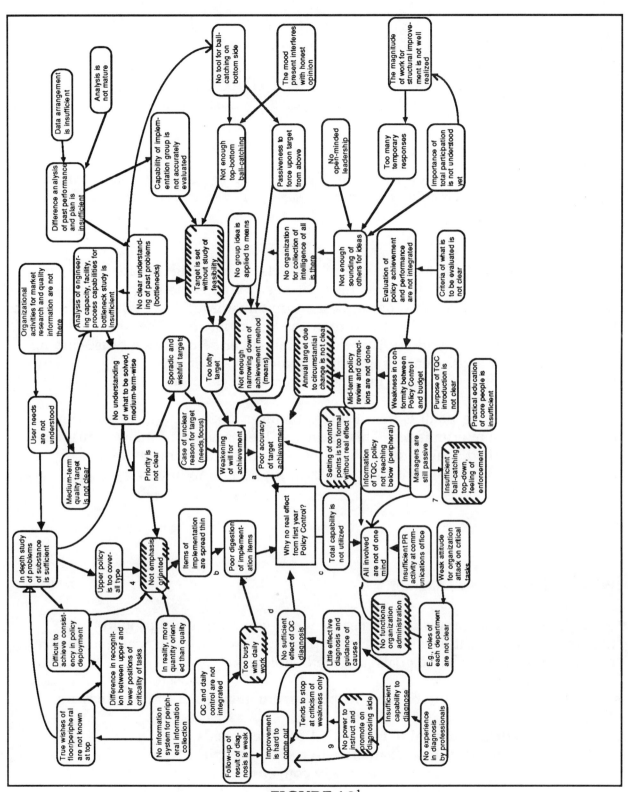

FIGURE 4.2[1]

---

[1]Nyatani, Yoshinobu. *Management by Policy to Promote TQC: Utilizing the New Tools of QC* (JUSE Press, Tokyo. 1982 Japanese) Figure 3-6.

SYSTEM FLOW/TREE DIAGRAM

Definition

This tool systematically maps out the full range of paths and tasks that needs to be accomplished in order to achieve a primary goal and every related sub-goal. In the original Japanese context, it describes the "methods" by which every "purpose" is to be achieved.

In many ways, the KJ Method and Interrelationship Digraph force the key issues to the surface. The questions then become, "What is the sequence of tasks that need to be completed in order to best address that issue?" or "What are all of the factors which contribute to the existence of the key problem?" The Tree Diagram is appropriate for either question. Therefore, it can either be used as a cause-finding problem solver or a task-generating planning tool. In either use it brings the process from a broad level of concern to the lowest practical level of detail.

Another strong point is that it forces the user to examine the logical link between all of the interim tasks. This addresses the tendency of many managers to jump from the broad goal to details without examining what needs to happen in order for successful implementation to occur. It also rapidly uncovers gaps in logic or planning.

When to Use the Tree Diagram

The Tree Diagram is indispensable when you require a thorough understanding of what needs to be accomplished, how it is to be achieved, and the relationships between these goals and methodologies.

It has been found to be most helpful in situations such as the following:

a. **When you need to translate ill-defined needs into operational characteristics.** For example, a Tree Diagram would be helpful in converting a desire to have an "easy to use VCR" into every product characteristic that would contribute to this goal. It would also identify which characteristics can presently be controlled.

b. **When you need to explore all the possible causes of a problem.** This use is closest to the Cause & Effect Diagram (Fishbone Chart). For example, it could be used to uncover all of the reasons why top management may not support a continuous improvement effort.

c. **When you need to identify the first task that must be accomplished when aiming for a broad organizational goal.** For example, the Tree Diagram would be very helpful to the coordinator of Quality Improvement Programs who wants to know what is already being accomplished and where the key gaps exist.

d. **When the issue under question has sufficient complexity and time available for solution.** For example, a Tree Diagram would not be particularly helpful for deciding how to deal with a product contamination problem that is shutting down your production line. It could be used to prevent it from reoccurring, but not in deciding on the stop-gap measures to be taken.

NOTE: In its most common usage the Tree Diagram conceptually resembles the Cause & Effect Diagrams. We have found it to be easier to interpret because of its clear, linear layout. It also seems to create fewer "loose ends" than the C&E.

Construction of a System Flow/Tree Diagram

It has been shown that these tools are most powerful when used in combination, but they are also very effective when applied singly. With this in mind the following are the most widely used steps:

1. Agree upon one statement that clearly and simply states the core issue, problem, or goal. This statement may or may not come from a KJ Chart or Interrelationship Digraph.

   Note: Unlike the KJ Method, the Tree Diagram becomes more effective as the issue is more clearly specified. This is important since the emphasis is on finding the logical and sequential links between ideas/tasks and not pure creativity.

2. Once the statement is agreed upon, the team must generate all of the possible tasks, methods, or causes related to that statement. These could follow three different formats:

   a. Use the cards from the KJ Chart as a foundation. For example, you might take the 10 - 20 cards that fall under one broad heading as a starting point.

   b. Brainstorm all of the possible tasks/methods/causes and record them on a flip chart. These ideas could then be placed on individual cards or rearranged on the flip chart.

   c. Brainstorm as in "b" but record directly onto cards for continued use.

      Note: When brainstorming, continue to apply to each idea the question "In order to achieve X, what must happen or exist?" Or "What has happened or what exists that causes X?"

3. Evaluate and code all of the ideas with the following code:

   O   Possible to carry out
   ^   Need more information to see if possible
   X   Impossible to carry out

   Note: Code an idea to be impossible only after very careful consideration. "Impossible" must not be equated with "we've never done it before".

4. Construct the actual Tree Diagram:

   a. Place the central goal/issue card to the left of a flip chart or table. (The remainder of the instructions will assume that cards are being used, but the same steps would apply if the chart is drawn directly on the flip chart.)

   b. Ask the question, "What method or task do we need to complete in order to accomplish this goal or purpose?"

Find the ideas on the cards or flip chart list that are most closely related to that statement. These may also be viewed as those tasks that are the closest in terms of sequence or cause & effect.

c. Place the ideas/tasks from "b" immediately to the right of the central issue card as you would in a family tree or organizational chart.

d. The ideas/tasks from "c" now become the focal point. In other words, the question from "b" is repeated and the remaining cards are again sorted to be placed to the right as the next row in the Tree. This process is repeated until all of the cards or recorded ideas are exhausted.
Note: If none of the cards answer the repeated question, create a new one and place it in the proper spot.

e. Review the entire Tree Diagram to ensure that there are no obvious gaps in sequence or logic. Check this by reviewing each path, starting at the most basic task to the extreme right. Ask of each idea/task, "If we do Y, will it lead to the accomplishment of this next idea/task?"

f. Review with other groups for relevant input and revise where needed.

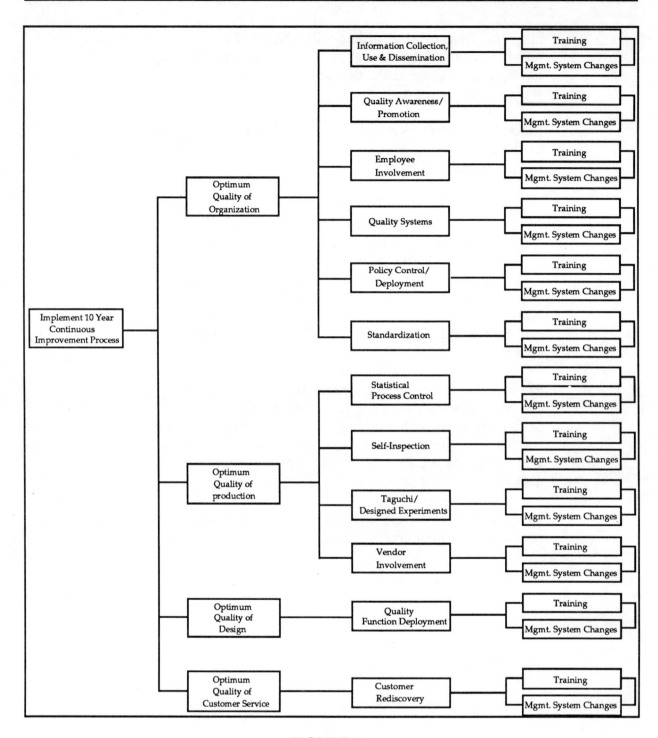

FIGURE 4.3

MATRIX DIAGRAM

Definition

This tool organizes large groups of characteristics, functions, and tasks in such a way that logical connecting points between each are graphically displayed. It also shows the importance of each connecting point relative to every other correlation.

Of the tools discussed thus far (KJ Method, Interrelationship Digraph, System Flow/Tree Diagram), the Matrix Diagram has enjoyed the widest use. It is based on the principle that whenever a series of items are placed in a line (horizontal) and another series of items are placed in a row (vertical) there will be intersecting points that indicate a relationship. Furthermore, the Matrix Diagram features highly visible symbols that indicate the strength of the relationship between the items that intersect at that point. The Matrix Diagram is very similar to the other tools in that new cumulative patterns of relationships emerge based on the interaction between individual items. Even in this most logical process, unforeseen patterns "just happen".

When to Use the Matrix Diagram

Because the Matrix Diagram has enjoyed the widest use of the New Tools, it has evolved into a number of forms. The key to successfully applying a Matrix Diagram is choosing the right format matrix for the situation. The following are the most commonly used matrix forms:

1. **L Shaped Matrix Diagram**
   This is the most basic form of Matrix Diagram. In the L shape, two interrelated groups of items are presented in line and row format. It is a simple two-dimensional representation that shows the intersection of related pairs of items as shown in Figure M-1. The Matrix Diagram may be used to display relationships between items in countless operational areas such as administration, manufacturing, personnel, R&D, etc.

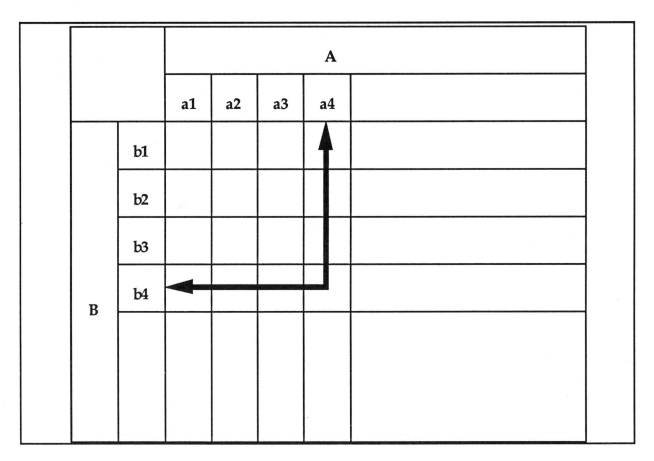

FIGURE M-1: L Shaped Matrix

In Chart M-1, the Matrix Diagram is used to identify all of the organizational tasks that need to be accomplished and how they should be allocated to individuals.
NOTE: It is doubly interesting if each person completes the matrix individually and then compares the coding with everyone in the work group.

Chapter 4: The Seven Management Tools

◎ Primary Responsibility
○ Secondary Responsibility   + Slightly More Emphasis
△ Communications/Need to Know

	Bob	Mike	Lee	Larry	Ann	Pat	Lynn	Board of Dir.	Other
Administration									
Payroll	◎						○	△	
Benefits	○	△	◎	△	△	△	△	○	
Office Systems	○	○	◎			◎	△		
Computer Programs	○	△	◎			○	○		
Courses									
Update Mailing List			○			◎	◎		
Select Courses to be Offered	◎	◎	◎			△	△	△ Deming	
Approve Course Content	◎	◎				△		△ Deming	○ Instructor
Prepare Brochures	○	○	○		◎	△			○ Instructor
Prepare Mailing			△			◎	○		
Hotel Arrangments	△	△	◎			△+	△		
Order Materials	△	△	◎			○	△		
Register People	△	△	△			◎	○		
Copy Materials	△	△	△			◎	○		
Prepare Packets	△	△	△			◎	○		
Room Set-Up	◎	◎	◎						
Post Receipts	△		◎			◎			
Prepare Bills									
New Course Development									
Market Research	○	△	△			△			
Implementing Deming	◎	◎	△	○		△	○		
TQC	◎	○	△						
Fundraising									
Annual Reports	○	○	○		◎	○	△		
Corporate Donations	◎	○	○	◎	○				
Committees									
Program Planning	◎	○	△			△			
Statistical Resources	○	◎	△	△	△	△			
TQC	◎	○	△	△	△	△			
Supplier Institute	○	○+		◎					

CHART M-1: L Shaped Matrix

Chart M-2 shows yet another application to an all-too-common problem: shipping problems. By brainstorming all of the possible reasons for shipping problems it is very clear that the "shipping" problem does not rest only with the shipping department. The matrix forces the participants to also develop the list of all related departments. The interrelationship between these two sets of items points to the pattern of responsibility for solution to the problems.

Problems \ Department/Individually/Function	Customer Service	Quality	Production	Scheduling	Process Engineering	Shipping	Design Engineering
Missing Parts	△		○			◎	
Does Not Meet Specs	△	○	◎		○		△
Wrong Parts	△	○		○		◎	
Mis-labeled	△		◎			○	
Defective Parts	△	○	◎		○		△
Arrived Late	△		○	◎		○	
Too Many	△		○			◎	
Too Few	△		○			◎	
Shipping Damage	△					◎	○
Wrong Part Ordered	◎						
Customer No Longer Needs Part	◎						
Cannot Process Part	△	○	○		◎		◎

◎ Primary Responsibility
○ Secondary Responsibility
△ Communications/Receive Reports

CHART M-2: L Shaped Matrix

2. **T Shaped Matrix**

The T Shaped Matrix is nothing more than the combination of two L Shaped Matrix diagrams. As shown in Figure M-2, the T Shaped Matrix is based on the premise that two separate sets of items are both related to a third set. Therefore, A items are somehow related to both B and C items.

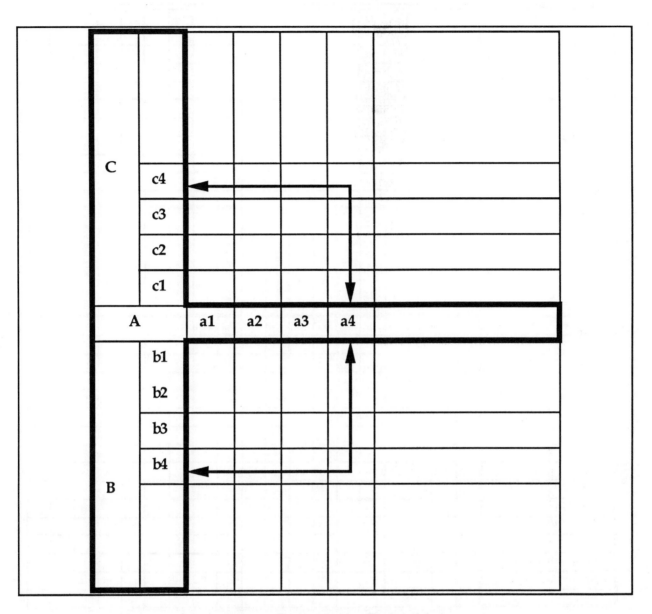

FIGURE M-2: T Shaped Matrix

Chart M-3 shows one application. In this case, it shows the relationship between a set of courses in a curriculum and two important sets of considerations: Who should do the training for each course? And, who would be the most appropriate attendees for each of the courses?

Who Trains?		Courses ... X = Full, O = Overview *Need to tailor to groups	SQC	7 Old Tools	7 New Tools	Reliability	Design Review	QC Basics	QCC Facilitator	Diagnostic Tools	Problem Solving	Communication Skills	Organize for Quality	Design of Experiment	Company Mission	Quality Planning	Just in Time	New Superv. Training	Comp. Tot. Q. Mgt. Syst.	Group Dynamics Skills	SQC Course/Execs.
	Human Resource Dept.																				
	Managers																				
	Operators																				
	Consultants																				
	Production operator																				
	Craft foremen																				
	GLSPC Coordinator																				
	Plant SPC Coordinator																				
	University																				
	Technology specialists																				
	Engineers																				
Who Attends?	Executives																				
	Top Mgmt.																				
	Middle Mgmt.																				
	Prod. Supervisors																				
	Supp. Func. Mgrs.																				
	Staff																				
	Marketing																				
	Sales																				
	Engineers																				
	Clerical																				
	Prod. Worker																				
	Qual. Professional																				
	Project Team																				
	Emp. Involv. Teams																				
	Suppliers																				
	Maintenance																				

CHART M-3: T-Matrix Diagram on Company-Wide Training

The T Shaped Matrix has also been widely used to develop new materials by simultaneously relating different, alternative materials to two sets of desirable properties.

3. **Y Shaped Matrix**

The Y Shaped Matrix simply allows the user to combine and compare three sets of items to each other. As shown in Figure M-3, it is clear that you can now determine the interaction between items in Group A with those in Group B, as well as Group B with Group C and Group C with Group A. This is invaluable when comparing product characteristics, etc.

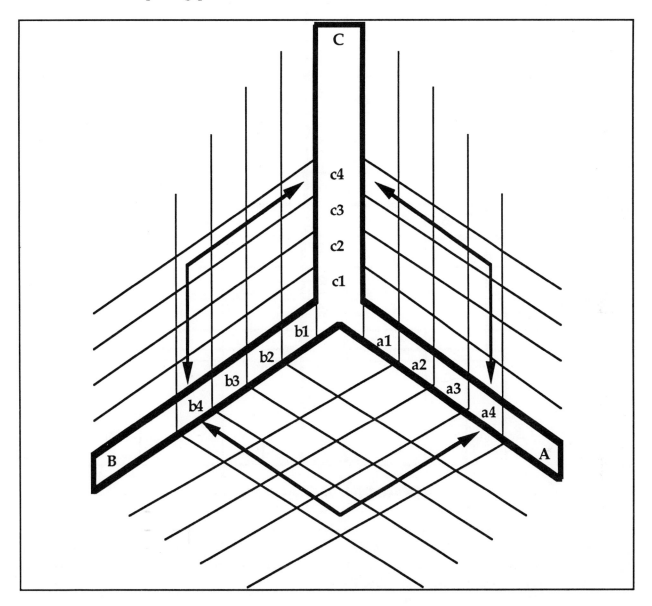

FIGURE M-3: Y Shaped Matrix

4. **X Shaped Matrix**

The X Shaped Matrix is a format that is rarely used. It shows the interaction between four sets of items. In Figure M-4, it graphically related A&B, B&C, C&D, and D&A. It is available but its use is not well documented.

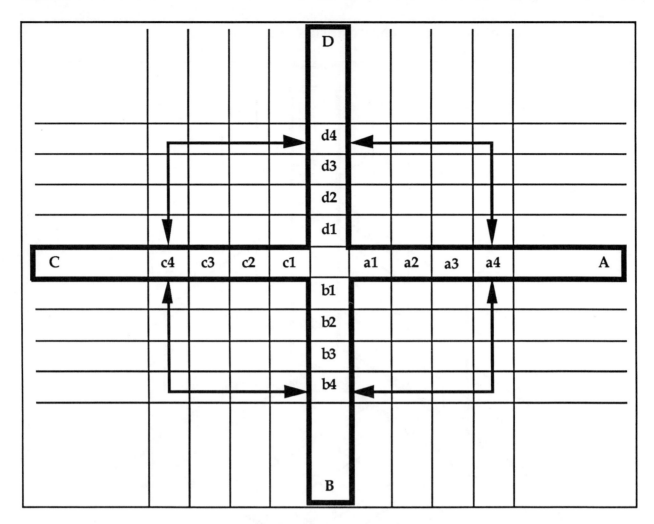

FIGURE M-4: X Shaped Matrix

5. **C Shaped Matrix**

   The C Shaped Matrix (or Cubic Type Matrix) makes it possible to visually represent the intersection of three interrelated sets of items.

   Other matrix diagram formats allow you to show the relationship between three or even four sets of items. In effect, however, they only compare two sets of items at a time with any connections to a third set only by inference. In other words, A is connected with B, and B is connected with C, so it can be inferred that A is related in some way to C.

   The advantage of the C Shaped Matrix is that it can graphically display the connection between A, B, and C directly as one converging point.

Chapter 4: The Seven Management Tools   p.4-23

Chart M-4 shows a C Shaped Matrix displaying the interaction between Layout, Software, and Hardware items. In this case, there is a strong connection between 4 under Layout (Select Software), 13 under Hardware (Measure Row Materials), and 9 under Software (Be Flexible Against Changes).

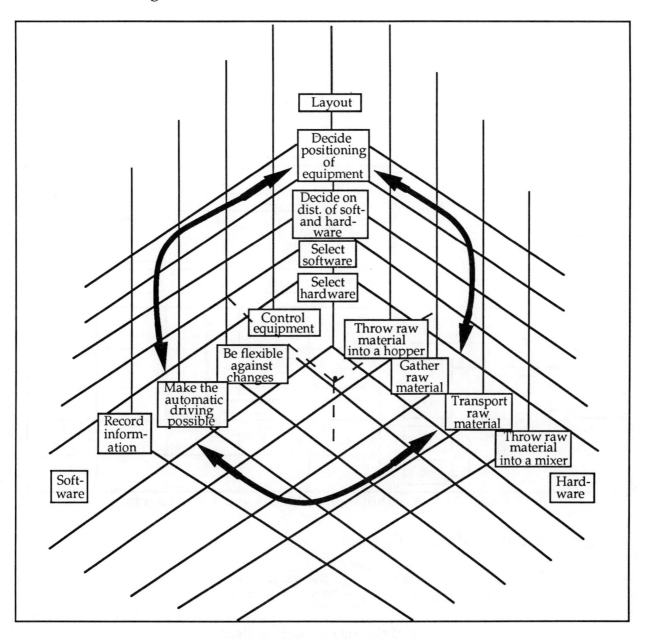

CHART M-4: C Shape Matrix

6. **Combination Matrix/Tree Diagram**

   Generating the most complete set of items possible is as important as selecting the right format matrix. The Tree Diagram is widely used to generate the tasks, ideas and/or characteristics that form one or more sides of the matrix.

   Figure M-5 shows two tree diagrams that have been merged into a simple L Shaped Matrix. Even more common than this is using a tree diagram to create a set of tasks to be accomplished (vertical axis of matrix) and merge them in an L Shaped Matrix with all of the departments/functions (horizontal axis). The degrees of responsibility for each task can then be clearly allocated and indicated.

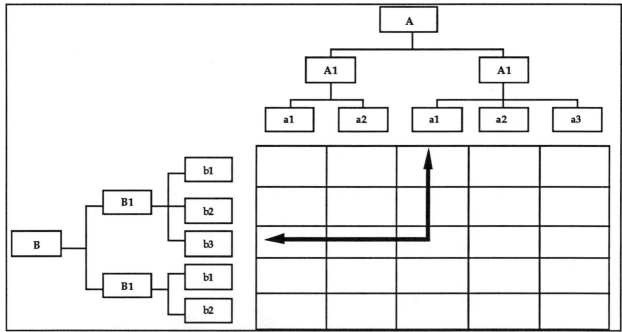

FIGURE M-5: Combination of Matrix and Tree Diagram

Construction of a Matrix Diagram

The process of constructing any of the various format Matrix Diagrams is very straightforward. It as as follows:

1. Generate the two, three, or four sets of items that will be compared in the appropriate matrix.
   NOTE: These often emerge from the last row of detail in a Tree Diagram. This is the most effective method, but the matrix has proven helpful when

based upon brainstormed items from a knowledge team.

2. Determine the proper matrix format.
   The choice of sets of items to compare is based on an educated guess and experience. It is trial and error. Don't be afraid to abandon or modify a line of reasoning.

3. Place the sets of items in such a way as to form the axes of the matrix.
   If these items come from one or more Tree Diagrams you can simply tape the cards (if used) on a flip chart pad. Otherwise you can simply record them directly on the pad. Finally, draw the lines which will form the boxes within which the appropriate relationship symbols will be placed (see step 4).

4. Decide on the relationship symbols to be used. The following are the most common, but use your imagination.

   - Function Responsibility Chart
     ◎ Primary Responsibility
     ○ Secondary Responsibility
     △ Should Receive Information
   - Quality Characteristics Chart
     A Most Critical
     B More Critical
     C Critical
   - Product Testing Chart
     • Test In Process
     O Test Scheduled
     X Test & Evaluation Possible

Note: Regardless of which symbols you choose to use, be sure to include a legend that prominently displays the relationship symbols and their meanings.

MATRIX DATA ANALYSIS

Definition

To arrange data displayed in a Matrix Diagram so that it can be more easily viewed and reveal the true strength of the relationship between variables.

When to Use Matrix Data Analysis

Matrix Data Analysis is primarily used for market research, planning and development of new products, and process analysis. It is used to determine the representative characteristics of each variable being examined. For example, what are the demographic characteristics of groups of people who like or dislike certain foods? What are the representative characteristics of a new cloth given an array of possible end uses?

Construction of a Matrix Data Analysis Chart

1. In order to find the "representative characteristics" of a product or consumer, use the "Principal Component Analysis Method". It is a formula that mathematically calculates the impact a factor has on the process.

2. Compare data among evaluation groups showing how much of the intergroup variation is due to a particular characteristic of that group (or combination thereof).

3. Calculate the cumulative contribution rates of the principle components to the overall ratings (e.g., sex, age, and occupation accounted for 75% of the variability in the rating).

4. Display the distribution of results graphically in a four quadrant graph.

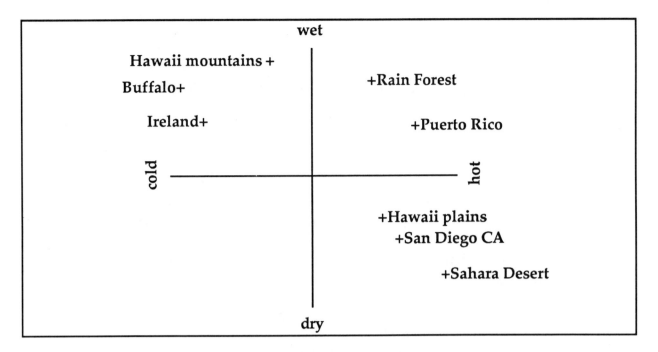

Illustration of Matrix Data Analysis Using Weather
FIGURE 4.4

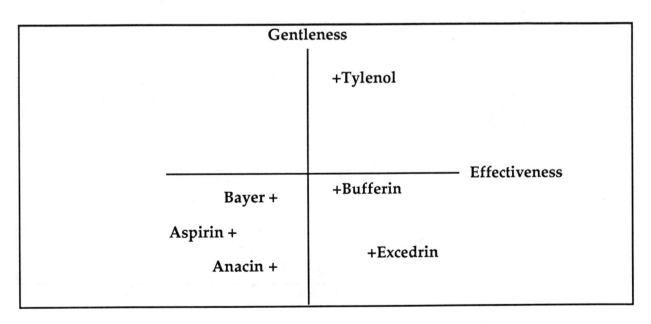

Illustration of Matrix Data Analysis Using Pain Relievers
FIGURE 4.5

Process Decision Program Chart (PDPC)

Definition

Process Decision Program Chart (PDPC) is a method which maps out every conceivable event and contingency that can occur when moving from a problem statement to possible solutions. This tool is used to plan each possible chain of events that needs to occur when the problem or goal is an unfamiliar one.

The underlying principle behind the PDPC is that the path toward virtually any goal is filled with uncertainty and an imperfect environment. If this weren't true, we would have a Deming "sequence" like the following:

PLAN ------------> DO

Reality makes the Deming Cycle a necessity.

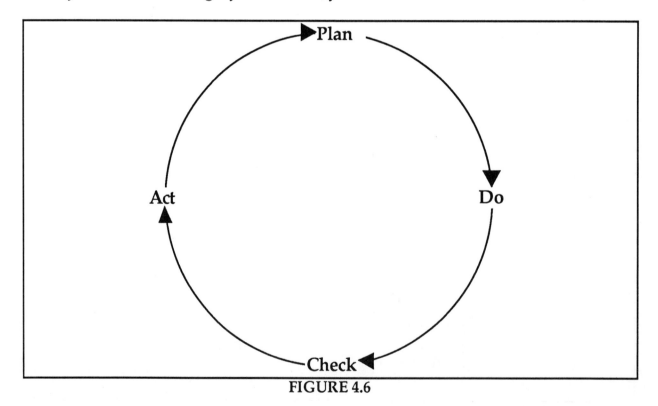

FIGURE 4.6

PDPC anticipates the unexpected and, in a sense, attempts to "short circuit" the cycle so that the "check" takes place during a dry-run of the process. The beauty of PDPC is that it not only tries to anticipate deviations, but also to develop countermeasures that will either:

a. prevent the deviation from occurring
b. be in place in case the deviation occurs

The first option is ideal in that it is truly preventive. However, we live in a world of limited resources. In allocating these resources we have to often "play the odds" as to the chance of X,Y, or Z happening. Given that fact, the next best thing is to have a contingency plan in place when a case that we were "betting against" occurs. PDPC provides a structure to go in either direction.

When to Use a PDPC

An ideal use of a PDPC would be as follows:

> A scientist's goal is to explore the core of the earth to determine its composition. Her plan is to dig a tunnel four miles deep. It's never been done, she doesn't know how long it will take, but she has to make a funding proposal. The questions are: How do you describe all of the possible paths to achieve this goal? How do you know what some of the obstacles will be? How can you prevent these possibilities from becoming realities? If obstacles do occur, how do you react in a timely way so as to avoid going back to "square one?" A cost estimate will be possible only if these questions have been answered. How can this be done systematically? Simple . . . . PDPC!

It is obviously not so simple, but it certainly provides a methodical structure that can prevent details from slipping between the cracks.
NOTE: PDPC is like the Tree Diagram in structure and aim since both deal with possible patterns of methods and plans. In the same vein, it is closely tied to methods in reliability engineering such as Failure Mode & Effect Analysis (FMEA) and Fault Tree Analysis (FTA).

The prime difference between these two formats is that FMEA starts from the smallest detail (sub-system) and assesses the probability of failure at any step. Also, it determines the cumulative impact on the end goal. FTA, on the other hand, starts with an undesirable result and then traces it back sequentially looking for the cause. PDPC is enjoying widespread use in particular because of the stress on product liability.

Construction of a Process Decision Program Chart (PDPC)

Even though PDPC is a methodical process it has few guidelines in terms of the process and finished product. The most important thing to keep in mind is that you must get to the point where deviations and contingencies are **clearly** indicated. This must be true at every level of detail in the chart.

Note 1: The source of the goal statement that starts the PDPC process often emerges from tools such as the KJ, Interrelationship Digraph, or even the Tree Diagram. As is true of all the other tools, PDPC can also be used effectively on its own.

Note 2: One word of caution. **EXPLOSION!** This is how some users have described PDPC. The creation of possible paths and countermeasures can multiply the complexity of the chart tremendously. Don't let it overwhelm you. Break the material into bite-size pieces, develop each piece, and then reassemble the final product.

The following seems to be the most workable approach:

a. Follow the instructions for the Tree Diagram through to the end.

b. Take one branch of the Tree Diagram (starting from the "purpose" in the row to the immediate right of the "ultimate goal/purpose") and ask the questions: What could go wrong at this step? or What other path could this step take?
   Note: It is easier if the items in that original branch are on cards so that they can be moved easily. This is important because you are inserting problems and countermeasures into an existing sequence.

c. Answer the questions in "b" by branching off the original path.

d. Off to the side of that step, list actions/countermeasures that could be taken. These are normally enclosed in "clouds" similar to cartoon captions and attached to that problem statement.

e. Continue the process until that original branch is exhausted.

f. Repeat "b" through "e" on the next most important tree branch, etc.

g. Assemble the individual branches into a final PDPC, review with the proper team of people, and adjust as needed.

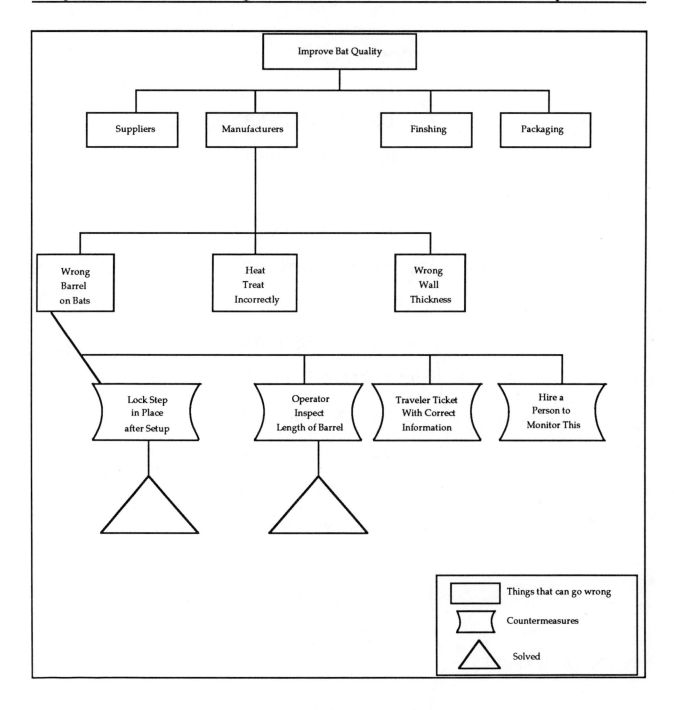

Process Decision Program Chart (PDPC)
FIGURE 4.7

PDPC of PLAN

STEPS
1. IDENTIFY NEEDS (CUSTOMER)
2. SELECT EQUIPMENT
3. TRAIN EMPLOYEES
4. INSTALL EQUIPMENT
5. HAPPY CUSTOMERS

PROBLEMS
3.0 TRAINING
3.1 NO DEMONSTRATION EQUIPMENT
3.2 EMPLOYEE APATHY
3.3 NOT ENOUGH TIME
3.4 NO MANUALS
3.5 TRAINING TOO LATE
3.6 TRAINING TOO EARLY

COUNTERMEASURES
3.2.0 EMPLOYEE APATHY
3.2.1 MORE MONEY
3.2.2 NO LAYOFFS
3.2.3 FIRE THEM
*3.2.4 BETTER MANAGEMENT
**3.2.5 BIG PICTURE

ARROW DIAGRAM

Definition

This tool is used to plan the most appropriate schedule for any task and to control the task effectively as it progresses. This tool is closely related to the CPM and PERT Diagram methods. It is used when the task at hand is a familiar one with sub-tasks that are of a known duration.

The Arrow Diagram is one tool that is certainly not Japanese. It is based on the Program Evaluation and Review Technique (PERT) which was developed in the U.S. after WWII to speed the development of the Polaris program. The Arrow Diagram removes some of the "black box magic" from the traditional PERT process. This is consistent with the general idea that the key to Japanese success is their ability to take previously available tools and make them accessible to the larger population. So, instead of industrial, manufacturing, and design engineers papering their walls with PERT charts (which they have done), they can be used as a daily tool throughout the organization.

When to Use the Arrow Diagram

The most important criterion (and perhaps the only meaningful one) is that the sub-tasks, their sequencing, and their duration must be well known. If this is not the case, then the construction of the Arrow Diagram can become a very frustrating experience. When the timing of the actual events is very different from the Arrow Diagram, people dismiss the Arrow Diagram as a nuisance, never to be used again. When there is a lack of process history, the PDPC is usually a much more helpful tool.

Note: Don't be afraid to admit that you don't know everything there is to know about a process. It is better to decide on the proper tasks and sequencing than to pretend that you have a handle on the scheduling dimension.

Obviously, there are many processes that do have a well documented history. Therefore, the Arrow Diagram has enjoyed widespread use in such areas as:

- New Product Development
- Construction Projects
- Marketing Plans
- Complex Negotiations

Construction of an Arrow Diagram

As usual, a successful process is based on having complete input from the right sources. It's possible that one person could have all of the needed information for structuring an Arrow Diagram, but it is highly unlikely. Therefore, assembling a team of the right people is usually the first step. This team would follow the steps listed below:

1. Generate and record all of the necessary tasks to complete the project.
   Note 1: It is strongly recommended that these tasks be written simply and clearly on cards (about business card size or slightly narrower). This is essential for moving the cards before the final lines and arrows are drawn. Expect to generate 50 - 100 such cards.
   Note 2: On the Job Cards be sure to write the task to be completed only in the top half of the card. Draw a line under the task, thereby dividing the card in half. The length of time to complete that task will be put in this space later.

2. Scatter the completed job cards and judge the inter-relationship between jobs. Determine the relationship among the cards (i.e., what **precedes, follows,** or is **simultaneous** to each job), and place them in the proper flow. Delete duplications and add new cards if jobs are overlooked.

3. Decide on the positions of the cards by finding the path with the most job cards in a series. Leave space between the cards so that "nodes" can be added later. (Nodes are the symbols that show the beginning and end of a task or event. Draw these in only when the various paths have been determined.)

4. Find the cards whose path parallels the first path, then the path that parallels that one, etc.

5. Once these paths are finalized, write in the nodes, number them, and add arrows between tasks in each path as well as those linking each path to the other.
   Note: Each task/job is made up of two nodes. The task that begins with node #1 and ends with node #2 is task 1, 2.

Chapter 4: The Seven Management Tools   p.4-35

6. Carefully study the number of days, hours, weeks, etc. for each task and complete each job card.

7. Based on #5, calculate the earliest and latest start time for each node.
   Note 1: This is critical if you are to calculate the Critical Path (as in CPM), which is the longest cumulative time that the tasks require. This is therefore the shortest time in which one could expect the final tasks to be completed.
   Note 2: The earliest and latest start times should be calculated using the following formulas:

1. a. **Earliest Node Time**
   Suppose there is a job that starts from the node i. "Earliest node time" is the day when the job can be started. It is expressed as t_i^E.

   b. **Latest Node Time**
   Suppose there is a job that ends at the node i. "Latest node time" is the day when the job must be finished. It is expressed as t_i^L. t_i^E and t_i^L will be written near the node.

2. **How to Calculate Earliest Node Time**

   Here is how to calculate earliest node time:

   a. Earliest node time of the starting point (node 1) in the arrow diagram is 0 (i.e., $t_j^E = 0$).

   b. When one job has a latter node j, its earliest node date t_i^E can be obtained using the following equation.
   $$t_j^E = t_i^E + D_{ij}$$
   Where t_i^E is earliest node time of starting node i of node j. D_{ij} is the necessary days of the job (i,j).

   c. When there are two or more jobs using the node j as their latter node, its earliest node time t_j^E can be obtained using the following equation.
   $$t_j^E = \max(t_i^E + D_{ij})$$

3. **How to Calculate Latest Node Time**

   a. Latest node time of the very last point (node n) in the arrow diagram has the same value as *earliest* node time of that node, i.e.,

   $$t^L_n = t^E_n$$

   b. When there is one job using node i as the starting node, its latest node time t^L_i can be obtained using the following equation.

   $$t^L_i = t^L_j - D_{ij}$$

   Where t^L_j is latest node time of following node j to the node i. D_{ij} is the necessary days for the job (i,j).

   c. When there are more than two jobs which use node i as a preceding node, its latest node time t^L_i can be obtained using the following equation.

   $$t^L_i = \min(t^L_j - D_{ij})$$

4. Determine the relationship between t^E_i and latest node time t^L_i at the same node.

   $$t^E_i = t^L_i$$

   There is the following relationship between earliest node time t^E_i and latest node time t^L_i at the same node.

   $$t^E_i = t^L_i$$

5. **Critical Path**

   Critical path is the longest path from the starting point to the finishing point on the arrow diagram. It is the series of jobs important to the schedule control. Critical path should be as shown in the following equation.

   $$t^E_i = t^L_i$$

   Along the path, t^E_i and t^L_i should be posted, and the path should be shown in a heavy directional line in the arrow diagram.

SYMBOLS

1. **Event, node:** these are the beginning and the finish of a job, and they are the connecting points to other jobs.

2. **Job, activity:** this is the element that needs a length of time.

3. **Dummy:** this is the element that shows the interrelationship between jobs, but needs no time.

4. **Numbers for nodes**: these are the numbers at the nodes or events, and they show which job is being referred to, or in what order it is placed.

p. 4-38 Chapter 4: The Seven Management Tools

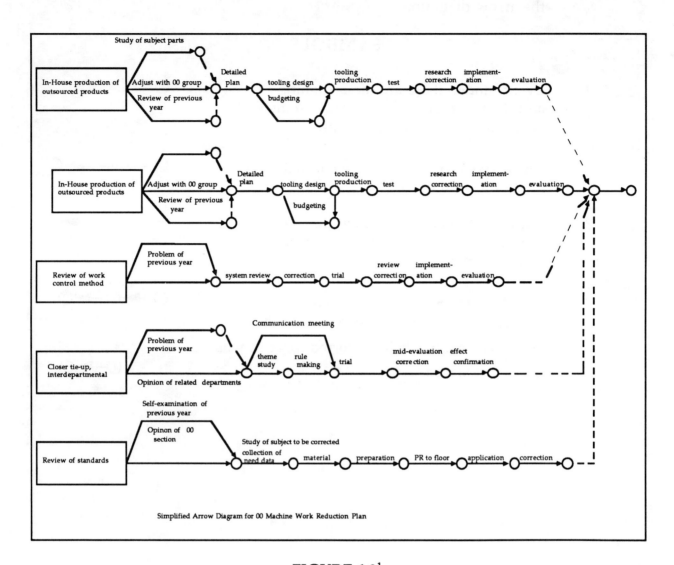

FIGURE 4.8[1]

---

[1]Nyatani, Ibid. Figure 3.32.

Chapter 4: The Seven Management Tools

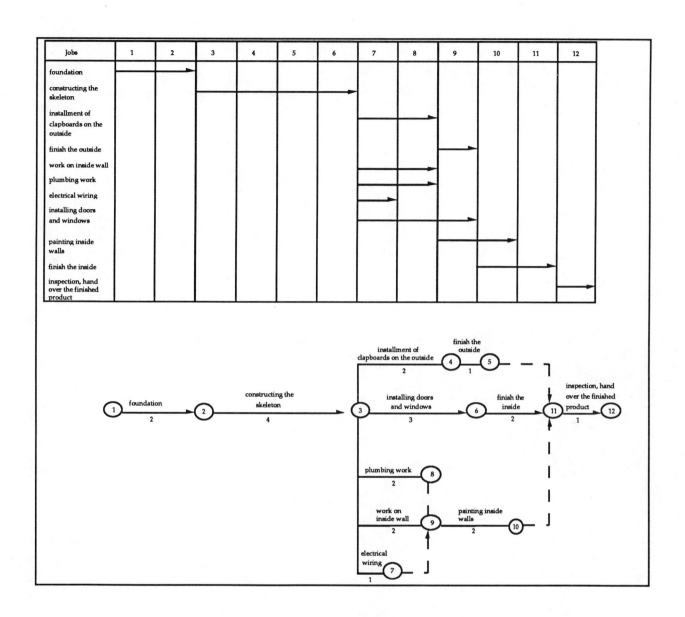

GANT CHART vs. ARROW DIAGRAM
FIGURE 4.9

Chapter 5
Target/Measures Matrix

As mentioned in Chapter 1, hoshin kanri may be translated as "target and means control". In the late 1970's, based on the successful use of matrices for QFD, Yoji Akao developed the target/means matrix to show the relationship between targets and means and to identify control items and control methods. The method used to generate the actual targets and means is based on the problem solving technology which was introduced in Chapter 3 and will be expanded on in Chapter 8. The targets and means are formatted on tree or system charts as described in Chapter 4, part 3. These two trees are then compared in a matrix as described in Chapter 4, part 4.

If we use our example of inadequate use of problem solving tools, we might generate a tree of targets that looks like this:

Reduce failures in problem solving actitivities	increase level of education
	increase problem solving skills
	work on the right problems
	hold the gains

FIGURE 5.1

Note: Figure 5.2 is in many respects a mirror image of the fish bone chart in Chapter 3, page 3.

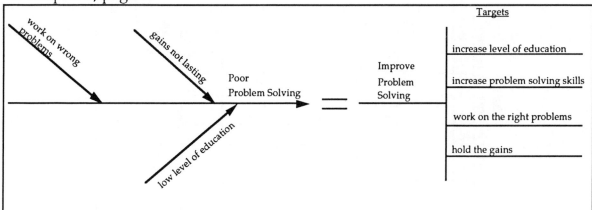

FIGURE 5.2

Note: "Increase problem solving tools" is a new item that is thought of and added.

Measures or means to achieve targets which were identified through problem solving methodology are arranged in a tree diagram in the same way.

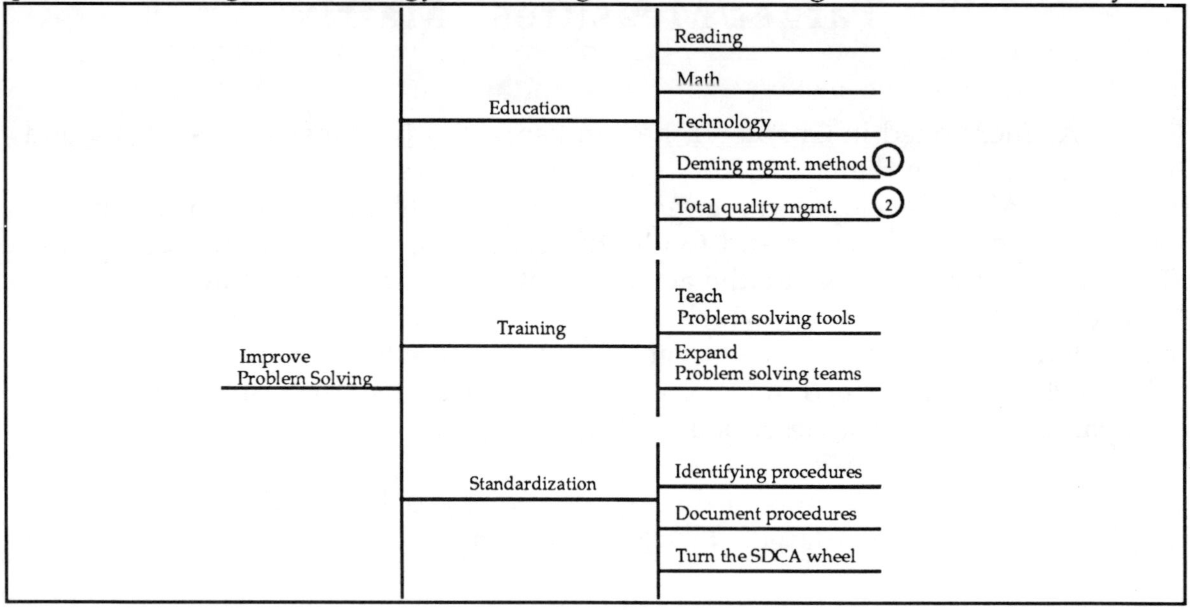

```
                                    ┌─ Reading
                                    ├─ Math
                      ┌─ Education ─┼─ Technology
                      │             ├─ Deming mgmt. method ①
                      │             └─ Total quality mgmt.  ②
                      │
Improve               │             ┌─ Teach
Problem Solving ──────┼─ Training ──┤  Problem solving tools
                      │             └─ Expand
                      │                Problem solving teams
                      │
                      │                  ┌─ Identifying procedures
                      └─ Standardization ┼─ Document procedures
                                         └─ Turn the SDCA wheel
```

FIGURE 5.3

These two trees are then combined into a matrix.

			WHAT **TARGETS** (Target deployment chart)			
			Reduce failures in problem solving activities			
			1.1 increase level of education about variation	1.2 increase problem solving skills	1.3 work on the right problems (fire prevention)	1.4 hold the gains after success
HOW **MEANS** (policy deployment chart)	1. Education	1.1 Teach reading skills				
		1.2 Teach math skills				
		1.3 Teach technology				
		1.4 Teach Deming Mgmt. Theory				
		1.5 Teach Total Quality Mgmt.				
	2. Training	2.1 Classroom—Teach problem solving skills				
		2.2 OJT—Practice problem solving skills				
	3. Standardization	3.1 Achieve breakthroughs				
		3.2 Identify procedures				
		3.3 Document procedures				
		3.4 Turn SDCA cycle				

FIGURE 5.4

Chapter 5: Target/Measures Matrix　　　　　　　　　　　　　　　p.5-3

The tree diagrams are arranged in blocks with the general items being whole numbers (1, 2, 3, etc.) and the sub-items being decimals. (1.1, 1.2,...2.1, 2.2, etc.)

The targets on the top side of the matrix are now compared with the means (measures, policies) on the left side of the chart.

			TARGETS (Target deployment chart)						
Policy: Improve problem solving			Reduce failures in problem solving activities						
⊙ Strong relationship ○ Some relationship △ Possible relationship			1.1 increase level of education about variation	1.2 increase problem solving skills	1.3 work on the right problems (fire prevention)	1.4 hold the gains after success	Registration #	Control Item	Control data
MEANS (policy deployment chart)		Teach							
	1. Education	1.1 Reading	△	△	△		M1.1	# at each reading level	histogram
		1.2 Math	○	○	△		M1.2	# at each math level	histogram
		1.3 Technology	○	○	○		M1.3	# attending tech. courses	linegraph
		1.4 Deming Mgmt. Theory	⊙	○	⊙	○	M1.4	# who have attended Deming course	linegraph
		1.5 Total Quality Management	⊙	⊙	⊙	⊙	M1.5	# course attendees	linegraph
	2. Training	2.1 Classroom—Teach problem solving skills	○	⊙	⊙	△	M2.1	# course attendees	linegraph
		2.2 OJT—Practice problem solving skills	⊙	⊙	⊙	△	M2.2	# problems solved with tools	linegraph
	3. Standardization	3.1 Achieve breakthroughs	○	○	○	△	M3.1	# suggestions implemented	linegraph
		3.2 Identify procedures	○	○	△	○	M3.2	# revisions in period of time	linegraph
		3.3 Document procedures	○	○	△	⊙	M3.3	# documents current	linegraph
		3.4 Turn SDCA cycle	⊙	⊙	○	⊙	M3.4	# procedures current	p-chart
		Ongoing audit procedures				⊙			
		Registration #	T1.1	T1.2	T1.3	T1.4			
		Control Item	# attending basic QMC	# using 7 QC Tools	% of time working on system	# of recurring problems			
		Control Data	line graph	line graph	line graph	line graph			

FIGURE 5.5

The double circle, circle, and triangle symbols are used in Japanese horse betting to represent win, place or show. They are used here to represent strong relationship, some relationship, or possible relationship. This is not in the narrow sense of the mathematical correlation but it can be made more precise by using it as a check sheet and recording evidences of relationships as they occur.

In this example, increased level of education about variation has a strong relationship with Deming's management theory, hence a double circle. Reading is seen as having a possible relationship with each target, hence a triangle.

Notice that Total Quality Management is the highest and one of the more general items. Care should be taken in constructing the tree diagrams when possibly similar levels of detail are compared to each other. Having a similar number of legs to the tree at each level helps, but these exercises and their tentative conclusions must be given the reality test: "Does it make sense?" "Is it consistent with what I know?"

These tools are structures for thinking and documenting considerations and analyses. They are not batch formulas into which you plug information and get automatically correct and perfect solutions. This is not to say they are not useful. Careful analysis in selecting the best goals and the best means to achieve them is essential to success as a manager and as a person.

Once the targets and means are clear, the next step is to identify the control items and control format. The identification numbers include a "T" for targets and an "M" for measures as well as the decimal number from the tree diagram.

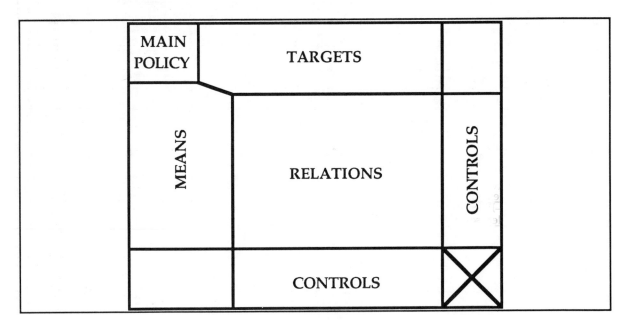

FIGURE 5.6

Chapter 5: Target/Measures Matrix

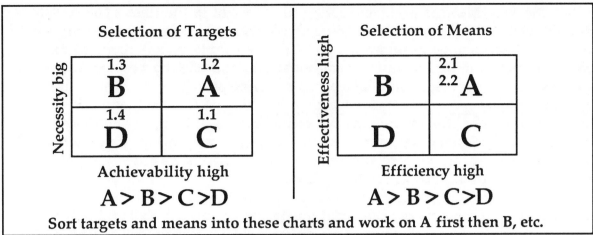

FIGURE 5.7

			TARGETS (Target deployment chart)						
			Reduce failures in problem solving activities						
Policy: Improve problem solving			1.1 increase level of education about variation	1.2 increase problem solving skills	1.3 work on the right problems (fire prevention)	1.4 hold the gains after success	Registration #	Control Item	Control data
MEANS (policy deployment chart)	1. Education	Teach							
		1.1 Reading	△	△	△		M1.1	# at each reading level	histogram
		1.2 Math	○	○	△		M1.2	# at each math level	histogram
		1.3 Technology	○	○	○		M1.3	# attending tech. courses	linegraph
		1.4 Deming Mgmt. Theory	◎	○	◎	○	M1.4	# who have attended Deming course	linegraph
		1.5 Total Quality Management	◎	◎	◎	◎	M1.5	# course attendees	linegraph
	2. Training	2.1 Classroom—Teach problem solving skills	○	◎	◎	△	M2.1	# course attendees	linegraph
		2.2 OJT—Practice problem solving skills	◎	◎	◎	△	M2.2	# problems solved with tools	linegraph
	3. Standardization	3.1 Achieve breakthroughs	○	○	○	△	M3.1	# suggestions implemented	linegraph
		3.2 Identify procedures	○	○	△	○	M3.2	# revisions in period of time	linegraph
		3.3 Document procedures	○	○	△	◎	M3.3	# documents current	linegraph
		3.4 Turn SDCA cycle	◎	◎	○	◎	M3.4	# procedures current	p-chart
		Registration #	T1.1	T1.2	T1.3	T1.4			
		Control Item	# attending basic QMC	# using 7 QC Tools	% of time working on system	# of recurring problems			
		Control Data	line graph	line graph	line graph	line graph			

FIGURE 5.8

Measure M-2-1 is problem solving tools taught in the class. The control item is the number of people attending the course. This is really a fundamental course that needs to be taken by all. In 1988 for example, Kodak reported 95% course participation by manufacturing personnel and 65% by non-manufacturing personnel. This progress could be tracked on a line graph.

At this point the reader may be concerned about the nature of these control items. Measures of courses attended, tools used, and even problems solved may appear repressive and bring to mind Dr. W. Edwards Deming's condemnation of arbitrary numerical goals.

And so it must be repeated that this target/means matrix is initially prepared for a manager to develop his or her own self-targets, self-means, and control items. These goals are not arbitrary and they are not for others. They are not for delegation. They are a tool for self-analysis and for setting personal goals. Figures 5.9 and 5.10 give examples of the use of this chart. In the first example, the division head's priority target is reduction of product claims.

The manager of production section #1 has established the target # Q1K as the maximum number of product claims per case. And he has identified improved, new product introduction. The means focus on reliability, storage (1-1-1), and better market information (1-1-2).

The control system for the target is the number of claims and the control item for the means is the level of improvement as measured with a P chart.

Figure 5.10 shows a target/means matrix from Kobayashi-Kosei.

Under the cost target, there is a target of at least 3050 cases more than the previous year (c1). This will be accomplished primarily by upgrading the people and their production methods. (6.2)

People must have a list of personal goals for success and align their goals with those of others. This matrix may be used for that alignment by combining several personal charts. Other alignment tools are described in the next two chapters.

Chapter 5: Target/Measures Matrix

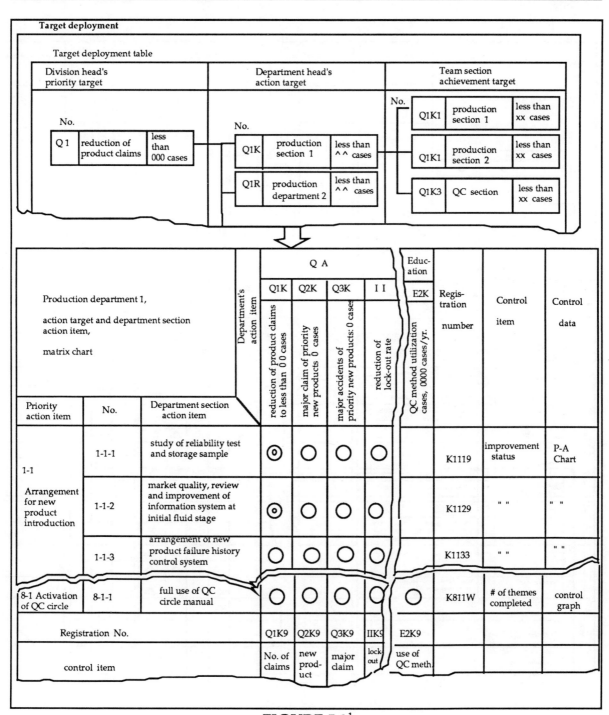

FIGURE 5.9[1]

---

[1] Akao, Yoji (Chairman of the Editing Committee) _Practical Applications of Management by Policy_ (Japan Standards Association, 1988, Japanese) p. 3-9c.

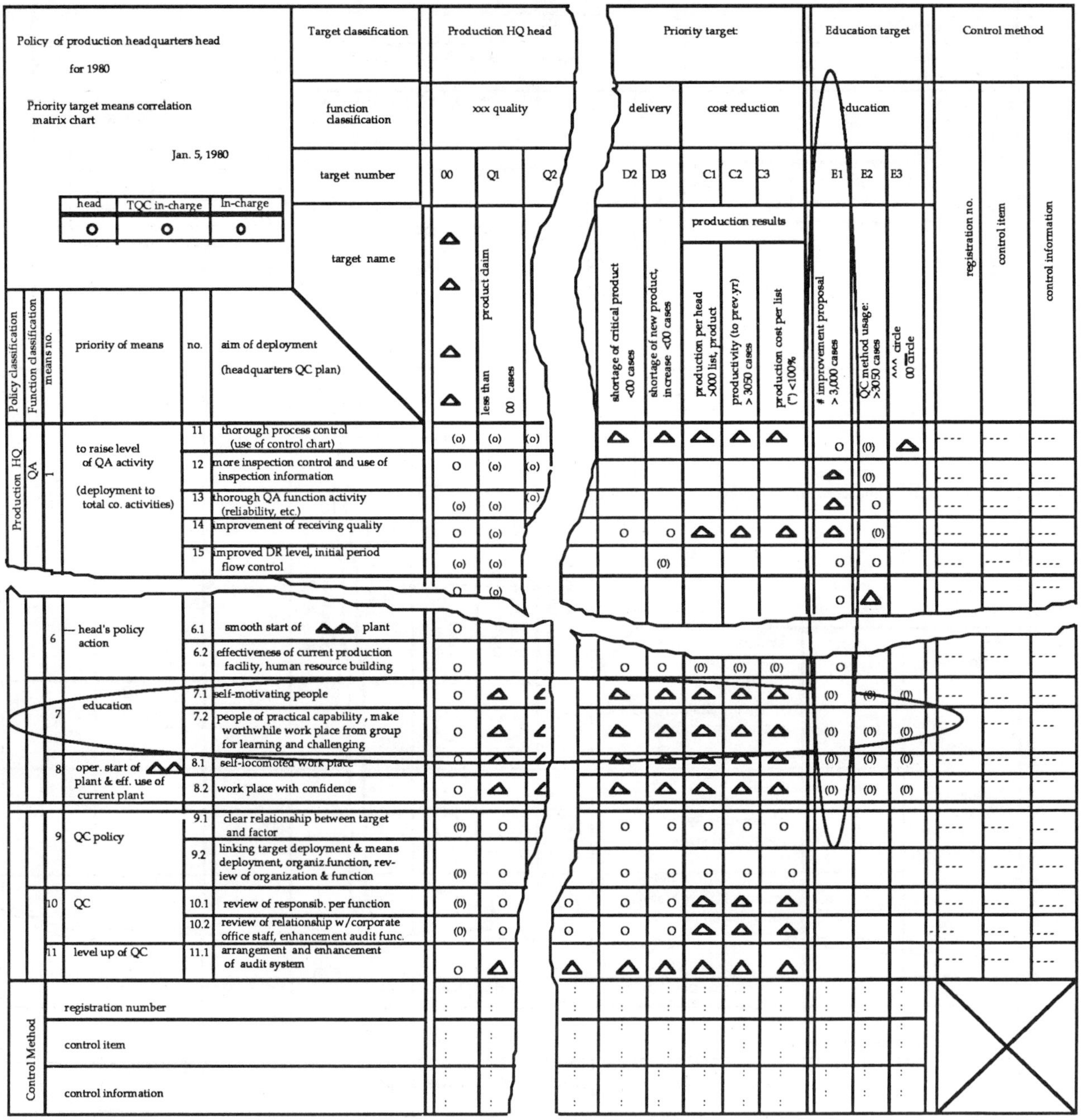

Chapter 5: Target/Measures Matrix

Example of Target Means Matrix Chart at Kobayashi-Kosei Production Hdqtrs.
(Onoguchi, Noda, Teisu-Hashi) 24

FIGURE 5.10[1]

---

[1]Nyatani, Ibid. Figure 3.11.

Chapter 6
The Flag System:
One Tool for Alignment

The flag system was developed by Komatsu in 1965 to ease the transition from Statistical Quality Control to Total Quality Control. The flag system is a tool that shows how key actions (means, policies) of each manager may be aligned to support key directions (targets) of the organization. This alignment is shown in the following modification of the tree/systems chart.

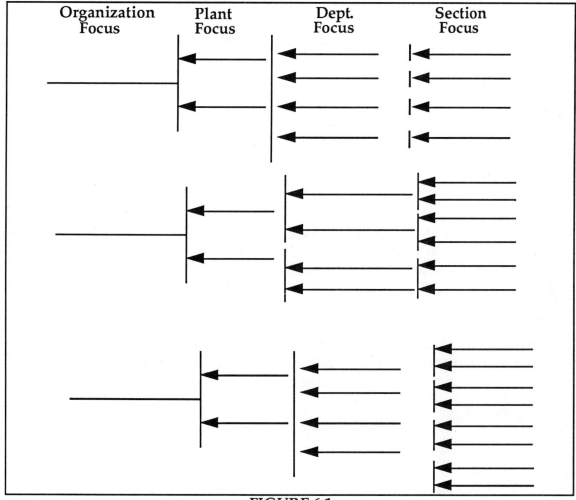

FIGURE 6.1
Schematic of Nesting or Alignment of Policies and Activities

Notice that the arrows are going from subordinate to the superior, thus showing a primary focus on goals that individual managers initiate and what each manager does.

The flag system is highly focused on the central aspect of policy deployment with monthly numerical targets and monthly numerical results. It is called the flag system because the arrangement of line graphs and connecting lines resembles vertical flags and horizontal flag poles.

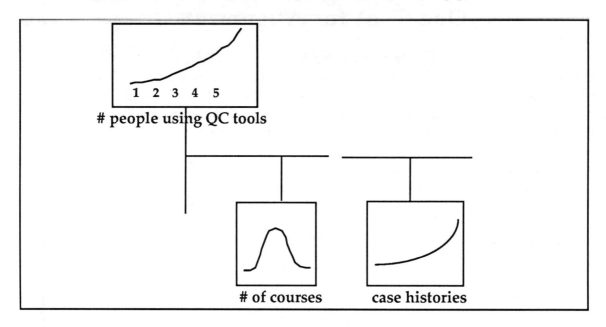

FIGURE 6.2

The selection of items to be measured is accomplished by the QC Tools of Chapter 3, The Seven Management Tools in Chapter 4, and the target/means matrix in Chapter 5. The continuous process work of hoshin planning focuses effort on where it is most needed. The steps are as follows:

Step 1: Identify a key problem to work on.
Step 2: Select appropriate targets.
Step 3: Select key factors to be measured to achieve targets.
Step 4: Hold an alignment meeting between subordinates and colleagues to compare flags.
Step 5: Fill in the target line.
Step 6: Develop an action plan.

Sequence:

1. More people using QC tools
2. Each department analyzes how to do that
3. Team meeting of managers to construct a flag system

Chapter 6: The Flag System: One Tool for Alignment

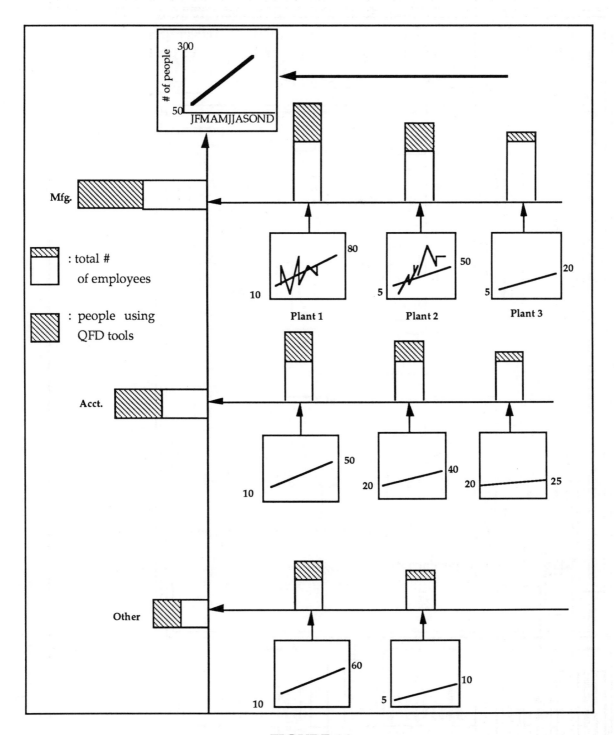

FIGURE 6.3

p. 6-4 Chapter 6: The Flag System: One Tool for Alignment

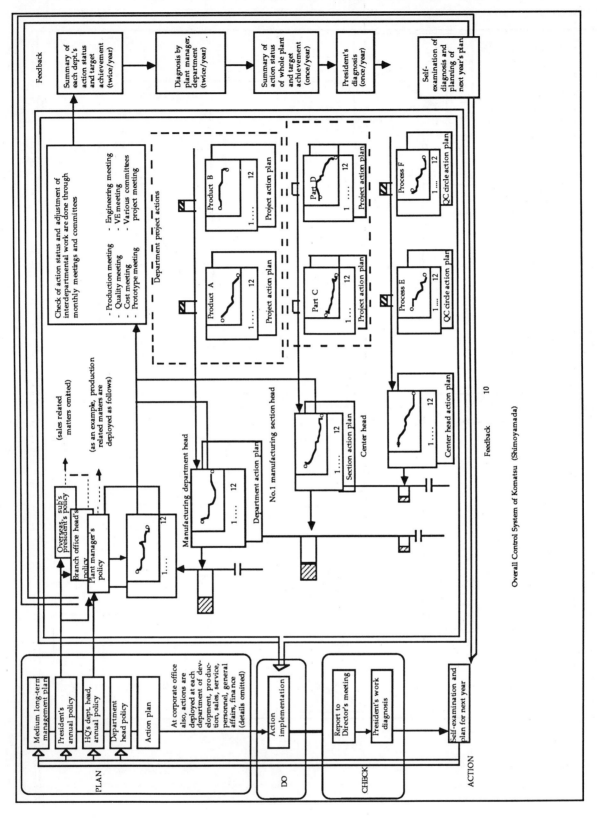

FIGURE 6.4[1]

[1]Nyatani, Ibid. Figure 3.7.

Chapter 6: The Flag System: One Tool for Alignment

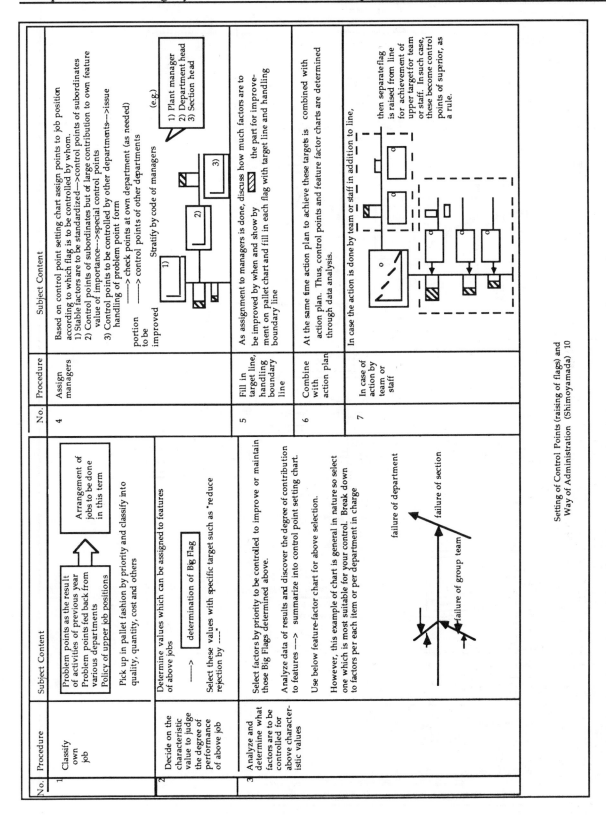

FIGURE 6.5[1]

---

[1]Nyatani, Ibid. Figure 3.8.

p. 6-6   Chapter 6: The Flag System: One Tool for Alignment

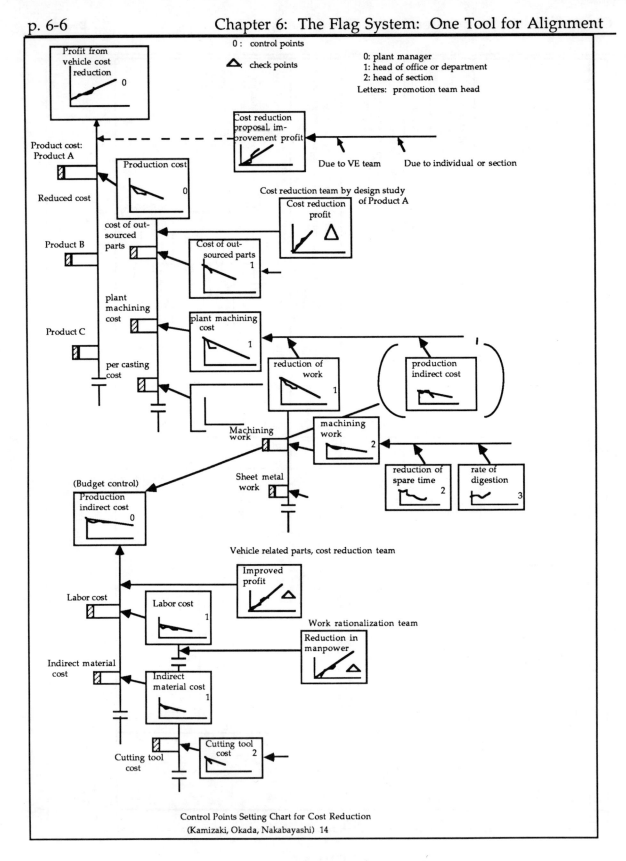

FIGURE 6.6[1]

---

[1]Nyatani, Ibid. Figure 3.9.

Chapter 7
Other Tools for Alignment

The flag system, which combines the pareto and fishbone charts, is an excellent way to align targets. Hoshin kanri focuses on the process and the means or policies which support the targets. Therefore, it would be a mistake to use only the flag system for alignment. Sometimes the targets and means are not related on a one-to-one basis. That is why the target/means matrix, developed by Yoji Akao, is so powerful. It enables the full analysis of targets and means even if they are disjointed. However, some find it helpful to use simpler tools to report the alignment after the planning work has been done. Some of the charts in this chapter may be useful in that regard.

Hewlett-Packard uses a cascading picture of objectives and strategies. (See Figure 7.1) This works well as long as they don't try to carry it too far down in the organization.

Deployment

Top management deploys the entity strategies and measures down through all appropriate levels of the entity.

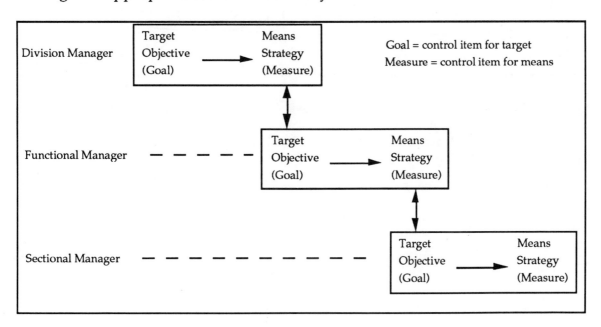

FIGURE 7.1

Kobayahi-Kosei uses a more detailed format which shows more of target and means interactions. (See Figure 7.2)

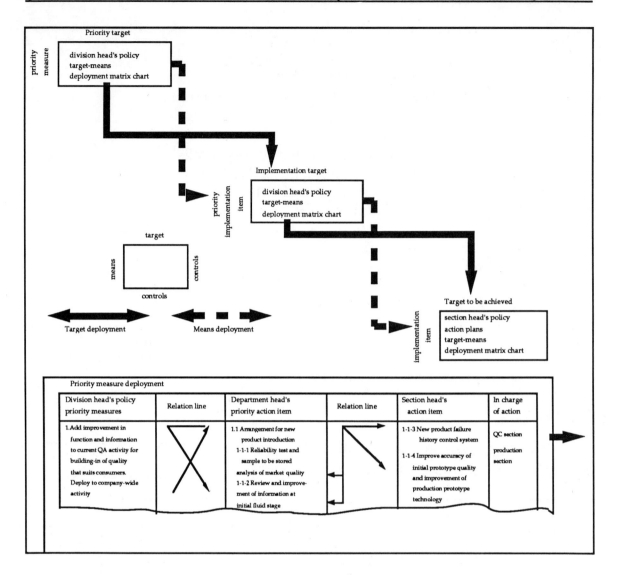

FIGURE 7.2[1]

Figure 7.3 shows one way to schematically portray the deployment of means alone.

---

[1]Akao, Ibid. Figure 3.10.

Chapter 7: Other Tools for Alignment

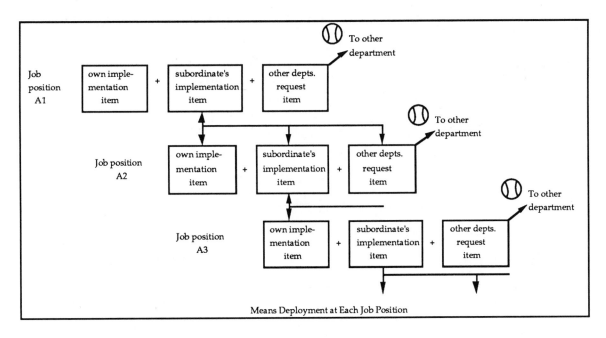

FIGURE 7.3[1]

The tree diagram can be used to show the deployment of targets. (See Figure 7.4)

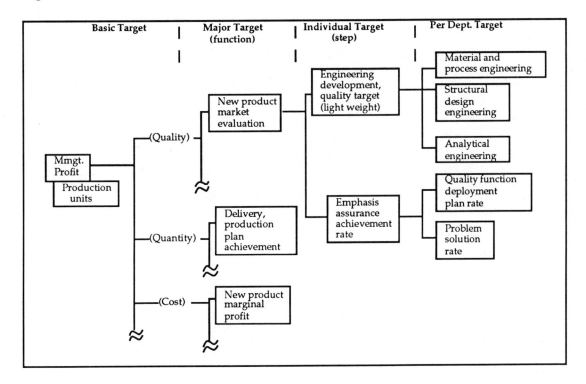

FIGURE 7.4[2]

---

[1]Nyatani, Ibid. Figure 3.28.
[2]Nyatani, Ibid. Figure 5.1.8.

The tree diagram can be used to show the deployment of means. (See Figure 7.5) The value of the tree diagram is found by adding the cumulative results from right to left. You can see if it will be enough to produce the hoped for results.

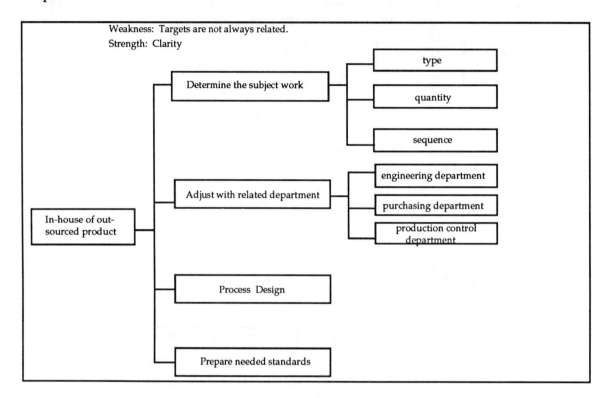

FIGURE 7.5

In the late 1970's, Toyota selected an improved hoshin planning system as a tool with which to dominate the automobile industry. One of their contributions was detailed integration of cross functional management with policy deployment. (See Figure 7.6) The arrows show the direction of influence. See Appendix B for more details on Toyota's advanced hoshin system.

Chapter 7: Other Tools for Alignment

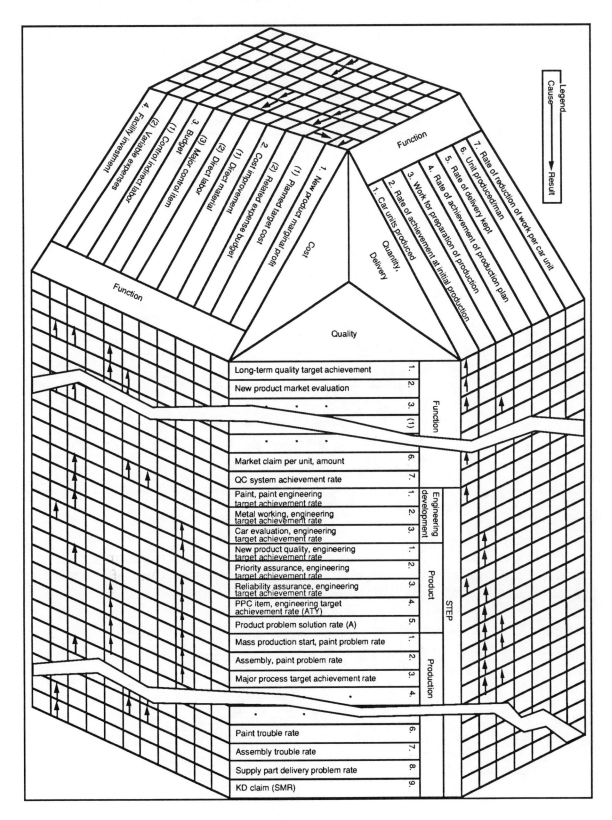

FIGURE 7.6[1]

---

[1]Nyatani, Ibid. 5.1.6.

Chapter 8
Hoshin Planning Phase 1
Process Management

PART 1: The Developmental Need for Process Management

Steps	1	2	3	4	5	6	Added Dimension
	Vision	1 Yr Plan	Deployment individual align	Execution (Process Mgmt)	Monthly Diagnosis	Annual Diagnosis	Developmental Learnings
(Ch. 8) Phase 1					●		Mgmt by facts
(Ch. 9) Phase 2			●	●	●		Self-diagnosis
(Ch. 10) Phase 3		●	●	●	●	●	Align -3 goals each
(Ch. 11) Phase 4	●	●	●	●	●	●	●	Understand new direction
Tools:	Aim	Plan	Do	Do	Do	Check/Act	Check/Act	
QC:								
Fishbone	●		●			●	●	
Pareto	●			●		●	●	
Line	●				●	●	●	
Flow	●		●		●	●	●	
Check Sheet	●					●	●	
Histogram	●					●	●	
Control	●				●	●		
7M: KJ	●	●						
ID	●	●					●	
Tree	●	●	●					
Matrix	●	●					●	
MDA								
PDPC				●				
Arrow				●				
Alignment Tools:								
Flag		●		●		●	●	
Target/Means				●				
Cascading T/M Tree				●				
QFD		●		●				

FIGURE 8.1

All organizations need to produce results to succeed. Auto companies need to design and build cars that people want and sell enough of them to cover the large developmental cost and provide a return for shareholders and R&D funds for the future. This is true whether it is automotive or electronics or hospital or insurance. These kind of results are basic to survival.

A focus on results only, however, can lead to great inefficiencies. People can use counter-productive measures to achieve the results. For example, a target can be set to cut defectives by 25%. A supervisor can then talk to his or her inspectors and say "You have been inspecting too tightly. You have been calling a lot of good product 'bad'. This is preventing our customers from getting goods they need. It is also giving us an unfair reputation as a poor quality organization." More than likely this will lead to looser inspection, some bad product being called good, and customers getting bad product, which will not be good for the company.[1] This may help the organization meet the target of lower reported defects but may harm the reputation of the organization by putting unacceptable product into the marketplace.

What is needed is to find out what is wrong in the operation and correct it. The way to do that is not to wait until the end of the operation, but to do the inspection at critical parts in the process. This helps pinpoint where the problems are in the process.

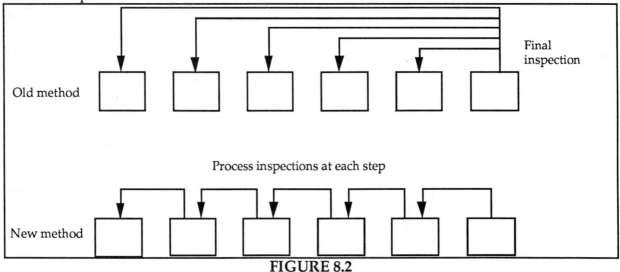
FIGURE 8.2

By finding the problems more quickly, this approach can reduce waste. It is also helpful because it identifies both isolated problems (special causes) and general problems (common causes). To succeed, it requires that everyone has knowledge of process control and problem solving skills.

A general problem solving approach can be used to identify problems and to find corrective action. There are a variety of tools that can be used, as illustrated below. (Excerpted from GOAL/QPC *Memory Jogger*.)

---

[1] An alternate approach to inspection is to re-inspect a sample of product from each worker and determine how many bad were called good and how many good were called bad and then work with inspectors to help them standardize their inspection.

Chapter 8: Hoshin Phase 1: Process Management

STEP	METHOD	
1. To decide which problem will be addressed first (or next)	• Flow Chart • Check Sheet • Pareto Chart	• Brainstorming • Nominal Group Technique
2. To arrive at a statement that describes the problem in terms of what it is specifically, where it occurs, when it happens, and its extent	• Check Sheet • Pareto Chart • Run Chart	• Histogram • Pie Chart • Stratification
3. To develop a complete picture of all the possible causes of the problem	• Check Sheet • Cause & Effect Diagram • Brainstorming	
4. To agree on the basic cause(s) of the problem	• Check Sheet • Pareto Chart • Scatter Diagram	• Brainstorming • Nominal Group Technique
5. To develop an effective and implementable solution and action plan	• Brainstorming • Force Field Analysis • Management Presentation	• Pie Chart • Add'l Bar Graphs
6. To implement the solution and establish needed monitoring procedures and charts	• Pareto Chart • Histogram • Control Chart	• Process Capability • Stratification

Another popular sequence of problem solving tools are those used by Ford Motor Company:

1. Define the problem
2. Identify the cause
3. Take corrective action (short-term)
4. Verify action taken
5. Implement the (long-term) correction
6. Verify the action taken
7. Standardize
8. Congratulate the team

In addition to the standard tools, the arrow diagram and PDPC (Process-Decision-Program) Chart can also be used to develop a long-term correction.

In the broader picture of hoshin planning, process management relates to the step of monthly diagnosis (step 5).

	Deal with problems	Purpose	Tools
5-1	Identify the problems	To describe where and when the problem occurs and its extent.	brainstorming, flow chart, check sheet, pareto, etc.
5-2	Identify the causes	To identify all possible causes and prime candidates.	check sheet, cause & effect, brainstorming
5-3	Take corrective action (short-term)	To try a solution to see if it will help.	modeling, force field analysis PDPC, arrow
5-4	Verify action taken	To see if the solution works.	pareto, histogram, control charts
5-5	Implement the (long-term) correction	To put the final corrective act in place.	PDPC, arrow
5-6	Verify the action taken	To see if long-term solution works.	pareto, histogram, control charts
5-7	Standardize	To make sure you hold gains.	quality characteristics, substitute quality characteristics, daily control
5-8	Congratulate the team	To recognize progress.	plaques, beer parties, etc.

FIGURE 8.3

Part 2: Tru-Save Example

The examples in this text are based on a supermarket example originally developed by Harry Artinian, previously of Ford[1], and guide lines originally

---

[1] Now VP of Quality for Colgate.

developed by Ed Baker, Director of Statistical Methods and Quality Planning at Ford. They are printed here with permission.

Tru-Save Example
by Harry Artinian

Comments for reflection
by Ed Baker

Example of use of tools:

Ashley Daniels gazed through the window of her second floor office overlooking the expansive main floor of one of the largest supermarkets in the Midwest. Her eyes became fixed on the front of the store where long lines of shoppers were beginning to form at each of twenty-four different check-out stations while cashiers hurriedly rang up overflowing carts of Saturday morning purchases. She just couldn't understand it. If there was one thing that personally irked her, that was it . . . and she was running out of ideas on how to improve the situation. She recalled how rising customer complaints had prompted her to initiate a "Cashier of the Month" campaign complete with Caribbean trips for the annual winner. The only thing that seemed to improve after 18 months was the travel agent's bank balance.

She then directed her managers to conduct daily inspections at different times to ensure cashiers were following procedure and not taking extra time by chatting too long with customers. She remembered feeling surprised to discover that cashiers were so busy that they hardly spoke to customers except to tell them what the purchase total was. Their perception of being "watched" had only resulted in lower morale.

As a last resort, and on the recommendation of her section managers, she authorized an increase in the check-out staff by four people. While there was a marginal improvement, over the last several weeks she began to notice (and this morning was no exception) that she now had long lines at 24 stations instead of 20. She believed she had an excellent staff but . . . what could she do to improve the situation? She decided to call a meeting of her five senior managers, the Head Cashier, and three of the store's best cashiers to develop a strategy that would yield permanent results.

On Monday afternoon Ms. Daniels met with her managers of Dairy, Frozen Foods, Health and Beauty, Regular Groceries, and General Merchandise sections, as well as Head Cashier Cathy Cash, and cashiers, Amy, Bonnie, and Pete. She began the meeting by stressing she wanted to improve the check-out system and would like to hear ideas on what they believed the inhibitors were to smooth customer flow at the check-out stations.

Cathy Cash began, "There are lots of items that have the wrong price sticker. The cashiers look to see if it should be marked down because it's dented or damaged but, if that's not the case, we have to call the manager for a price check."

Bonnie added, "Yes, and that's assuming the item had a price sticker in the first place. There are a lot of times on every shift when I get stuff that doesn't have one. Again, we have to call the manager for a price check. The stock boys don't seem too worried about making sure the stickers stay on."

Amy interjected, "And it's not just the stock boys. The secretaries that type the "Look-up Sheets" make mistakes on the codes and they're hard to read even when they are right. You remember how the sheets work . . . when I pick an item from the belt, the first decision I have to make is whether it's one of the over 200 high-turnover items that have no sticker but have assigned codes on the sheets instead. If it is, I look up the code and enter it, and the register assigns the price automatically. I know it saves the stock boys time because they don't have to re-sticker items that move fast and change price quickly, but it adds time at the front, especially when you're trying to be careful and find the right code. And you find yourself checking the register display after it's entered so you don't end up charging the price of a case of Coke against a jar of baby food. Customers get pretty mad about that and I can't blame them."

"I get a lot of customers who think they can write a check without proper identification", Pete was speaking up now. "And it depends which manager is on duty whether an I.D. is needed. If the manager knows the customer, everything is O.K. But it puts us in the middle and is real frustrating when a check is denied today but another customer with the same problem is accepted tomorrow. It just takes a lot of extra time when customers don't know our check cashing policy and we have to go through all the verification steps over and over during the shift while there are people lined up. And I just can't work any faster."

Ashley Daniels adjourned the meeting and returned to her office with the new information. Use the principles of process management to help her improve the quality of supermarket service.

<u>Principles</u> by Ed Baker, Office of Statistics and Quality Planning, Ford Motor Company

- Everyone manages a process.

- The business is a network of interdependent processing systems linked in supplier/customer relationships.

- Systems can be defined from macro to micro levels for analysis.

Chapter 8: Hoshin Phase 1: Process Management   p. 8-7

- Boundaries are defined by process flow.

- Responsibilities may have cross functional boundaries. Therefore, processing system ownership must be made explicit.

- Process control requires measurement and feedback to the process manager.

- Targets, objectives, and measures should help the process manager (supplier) meet customer requirements.

- Process improvement requires communication between customer and supplier.

- Improvement is process, not outcome oriented. Outcomes are measured over time to understand performance patterns in order to improve the process.

- Value is added when the customer is helped.

Using the eight Ford problem solving steps, let's look at the Tru-Save example.

Step 5-1: Identify Problems[1]

Example #1: Check Sheet Customer Complaints

lines too long   ⟋⟋⟋⟋ ⟋⟋⟋⟋ ⟋⟋⟋⟋ ⟋⟋⟋⟋ |||
not enough wine selection ||
no Sam Adams beer   ⟋⟋⟋⟋ ||
bags filled too full ||||
not enough quick foods |||
Nintendo games out of stock   ⟋⟋⟋⟋
chemicals in fruit |||
shelves too high ||
food gets moved too often |

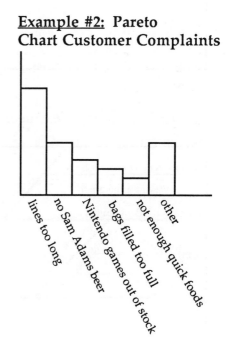

Example #2: Pareto Chart Customer Complaints

(bars labeled: lines too long, no Sam Adams beer, Nintendo games out of stock, bags filled too full, not enough quick foods, other)

---

[1]This step and the following steps are called 5-1, 5-2, etc. because they refer to step 5 (monthly diagnosis) on the hoshin planning master plan. They will be combined with the other major steps in the next three chapters.

Example #3: Flow Chart

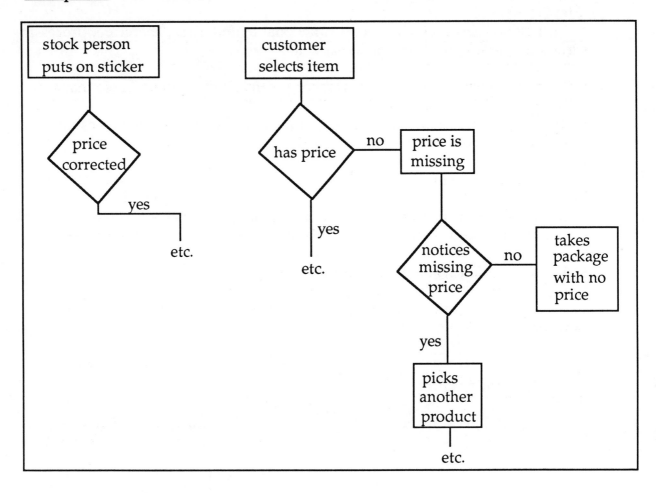

Step 5-2: Identify the causes (cause and effect diagram)

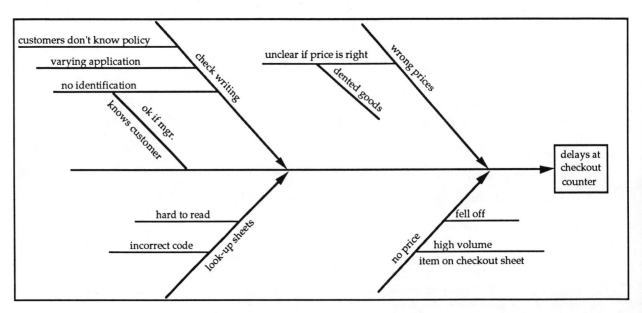

Step 5-2 Example 2: Checksheet

Reasons for delays	
wrong price	⊞⊞ ⊞⊞ ||
unclear if price is right (e.g., dented)	⊞⊞ |||
no price (fell off)	⊞⊞ ⊞⊞ ⊞⊞ ⊞⊞ ⊞⊞
look-up sheets hard to read	⊞⊞	
look-up sheets incorrect	⊞⊞ ⊞⊞ |||
check writing policy not known	⊞⊞ ||
no identification	|||

Step 5-2 Example 3: Pareto

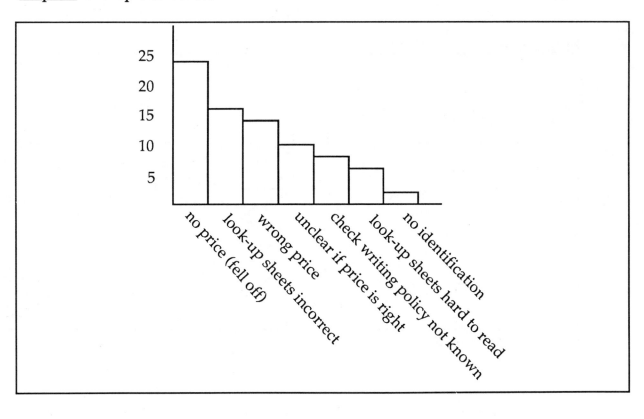

Step 5-3: Take corrective action short-term. Tell stock people to be more careful.

Step 5-4: Verify action taken

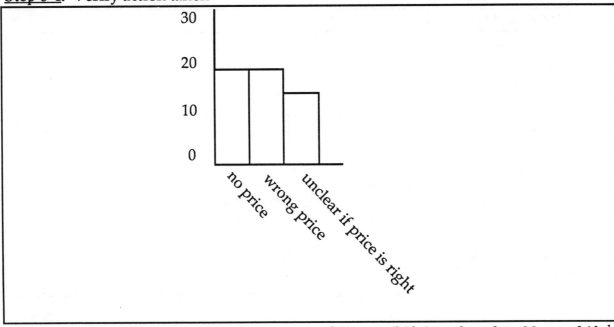

Action is minimally successful. Occurrence of 25 per shift is reduced to 20 per shift.[1]

Step 5-5: Implement long-term solution
Checkers taught how to affix labels, proper position of label gun, etc.

Step 5-6: Results for Month

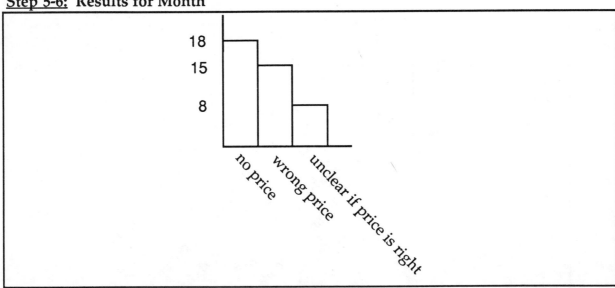

The improvement, though slight, is holding.

---

[1] In reality, the PDPC cycle might continually be turned until acceptable results are achieved. In this case, some of the breakthroughs are being saved for a later chapter.

Step 5-7: Standardize
1. Affixing price tags has become part of the training for all checkers.
2. Each new employee must watch a video on correct affixing of price tags.
3. Checkers are monitored on a random basis to see if they are affixing price tags properly.

Step 5-8: Congratulate the team

During the weekly employee meeting, those involved in developing the training materials for affixing labels are congratulated as well as those who identified the labeling method as important. All are encouraged to think of additional ways of addressing the label problem.

PART 3: Dr. Deming's 14 Points

The steps and tools in process management if applied alone will lead to some improvement. But there are many management practices in the U.S. that can greatly inhibit or kill the benefits of hoshin planning. These include short-term focus, performance appraisals, etc.

Dr. W. Edwards Deming has described these items as the deadly diseases of American management. He has also identified 14 Points for Management which provide a basis for setting the stage for successful process management. It might be helpful to review each of these in some detail.

1. Constancy of Purpose

An organization should always think long-term rather than concern itself only with the quarterly dividend or the monthly quota. The company must focus on making a quality product, customer demand today, and what customer demand will be five years from now. The aim is to be competitive, to stay in business, and to provide jobs.

2. Adopt a New Philosophy

In today's new economic age, we cannot live with the commonly accepted style of American management, nor with the commonly accepted levels of delays, mistakes, and defective product. People used to think that management's job was meeting specifications. Today 'good enough' is no longer good enough. Management must go beyond meeting specifications and produce a good, quality product.

3. Cease Dependence on Mass Inspection

Dr. Deming is not suggesting that inspection be entirely eliminated, but that companies focus on process and perform inspection for process control and critical parts. One hundred percent final inspection is too late and too expensive. The idea should be 'do it right the first time'.

4. Cease Doing Business on Price Tag Alone

Rather than awarding business on the basis of the price tag, organizations must strive to minimize total cost. This point reflects the changing role of purchasing. The old role of purchasing was to obtain product at the lowest unit cost possible. The problem with that is in order to achieve such a low cost, the manufacturer probably cut corners by using sub-standard materials or under-skilled, inexpensive labor.

The new role of purchasing is to more intensely communicate with suppliers. The customer must clearly define what he wants and the supplier must clearly explain how his product is made.

5. Continual Improvement of Processes

Management controls the system and has the responsibility to improve the system, whereas workers simply work within the system. Management is 85-90% to blame for what goes wrong but has that same percent opportunity to improve the system.

Processes are not just related to manufacturing. Process should be thought of as everything done in an organization from answering the phone to the most complex task. To improve the process, an organization must have a step-by-step plan of what will be done differently to make the improvement. Management must train people, update the systems, and provide workers with the right tools.

6. Institute Training on the Job

Many companies do not provide a lot of training. Many workers think their job is to do what the supervisor tells them to do. The problem is that at the beginning of the month, people are told to do quality work, and at the end of the month, workers are told to hurry because the product has to be shipped. The worker's job changes and contradicting directions from supervisors leads to frustration.

7. Institute Leadership

This point concerns the role of supervisors. Around 1900, Frederick Taylor concluded that most supervisors were unable to figure out the best way to do a job, and so an engineer should plan the job and then the supervisor can carry out the engineer's plan. Over time, however, the education level of supervisors improved, but the the system wasn't changed. The system must keep up with changes and today's supervisors must act as leaders.

Chapter 8: Hoshin Phase 1: Process Management   p. 8-13

8. <u>Drive Out Fear</u>

This can be the most difficult of the 14 Points because it is the most pervasive. Performance appraisals, goals, sometimes even threats are used to push workers to produce high quality at a low cost. Each of these methods contribute to fear in an organization and if there is fear, workers will not perform well. Workers must feel secure to be able to contribute. Fear is a barrier. It makes people afraid to report problems and afraid of approaching managers.

9. <u>Break Down Barriers Between Departments</u>

If people have been rewarded and conditioned to pursue their own goals, this step is difficult to accomplish. One way to break down barriers between departments is to establish project teams which are entities within an organization that bring the right people together with the right tools to work on the right problems. Instead of individual problems being the focus, the problem becomes the focus of the group.

10. <u>Eliminate Slogans, Exhortations, and Targets</u>

Has a poster or slogan ever helped you do a better job? Slogans and targets often frustrate people because they bring the contradictions of management into better focus. "Quality is Job One" is true for workers at the beginning of the month but not at the end when the supervisors are rushing to meet their quotas.

Another problem with posters is what Deming calls the 85/15 rule. Management has 85% control over the system and workers have 15% control, yet posters urge workers to control things. The bulk of the causes of low quality and low productivity belong to the system and thus lie beyond the power of the work force.

11. <u>Eliminate Numerical Quotas</u>

Managers use stretch goals to maximize worker performance, but Deming believes such goals are actually a floor that cause people to put forth the minimal effort needed to reach the goal.

If someone is given a goal to make 100 parts in an hour and he only makes 85, the focus should not be on the fact that only 85 were made. It should be on why only 85 were made. What kept the person from making more - the materials? - improper training? A manager should train the worker, tell him what he wants, and provide him with the right tools to do it.

12. <u>Allow Pride in Workmanship</u>

People want to be proud of their work. Certain management techniques prevent pride. Performance appraisals, for example, prevent pride in workmanship because they often pit one worker against another. The worker becomes concerned not just with what he's doing but also with what other workers are doing.

13. <u>Institute a Program of Self-Improvement</u>

This point should apply to everyone at all levels of the organization. It suggests that people take courses, get involved in community activities, and pursue new hobbies, thus gaining a wider perspective on things. The idea is to encourage the workers to broaden their horizons so that they can bring more to their jobs, have a wider vision. The person will become richer and fuller and so will be able to contribute more to the continuous improvement effort.

14. <u>Do It</u>

Each manager must internalize the 14 Points if the company is going to be successful. Management has to not only train people to do continuous improvement but also provide the environment for people to do it in. Management must create the capability to transform the organization into a competitive company and take part in the effort, not just get on a soapbox and preach to the workers. Continuous improvement can't just be a slogan on the wall, it must be internalized and carried out by every person every day for the transformation to take place.

Chapter 9
Hoshin Planning, Phase 2
Management Self-Diagnosis

PART 1: The Developmental Need for Self-Diagnosis

The first level of planning activity is setting targets and working on them. The second level of planning is process management, which involves all employees working on the system. (Described in Chapter 8) The next step is self-diagnosis; to enable each manager to use problem solving activity to see why they are succeeding and or failing to meet targets as a basis for continuous improvement.

Self-diagnosis for improvement is not an automatic reflex action in most organizations. The more frequent mode of acting is CYA (Cover Your Behind). Again and again review meetings are held in which the first ten minutes are spent realizing that the target was not met and the next hour and one half is spent discussing how to justify to management that the target could not be met.

The shift to the higher level of planning represented by manager self-diagnosis requires a quality culture (as represented by Deming's 14 Points in the previous chapter). Managers need to be able to fail and learn from their failures. Systems that set arbitrary numerical goals tied to performance appraisals will not be able to reach this level. Organizations based on fear and top-down management only will not be able to reach this level.

The setting of personal targets is a key to phase 2. An individual manager must set personal goals with no fear of reprisal if he or she does not meet them. This marks a significant departure from organizations where the boss sets the goals for the subordinate or where boss and subordinate negotiate goals.

In phase 2, problem solving tools (step 5) are used to understand why an individual goal was or was not met. Each month the results are reviewed. Did I make the target I set for myself or did I miss it? The focus then is on the process. What helped me make the target? What kept me from making the target?

As indicated in Figure 9.1, phase 2 includes step 3a: setting an individual target, step 4: execution and step 5: monthly diagnosis.

Steps	1	2	3		4	5	6	Added Dimension
	Vision	1 Yr Plan	Deployment		Execution (Process Mgmt)	Monthly Diagnosis	Annual Diagnosis	Developmental Learnings
			individual	align				
(Ch. 8) Phase 1						●		Mgmt by facts
(Ch. 9) Phase 2			●		●	●		Self-diagnosis
(Ch. 10) Phase 3		●	●	●	●	●		Align -3 goals each
(Ch. 11) Phase 4	●	●	●	●	●	●	●	Understand new direction
Tools:	Aim	Plan	Do	Do	Do	Check/Act	Check/Act	
QC:								
Fishbone	●		●			●	●	
Pareto	●			●		●		
Line	●				●	●	●	
Flow	●		●			●	●	
Check Sheet	●					●		
Histogram	●					●		
Control	●				●	●		
7M: KJ	●	●						
ID	●	●					●	
Tree	●	●	●					
Matrix	●	●					●	
MDA								
PDPC					●			
Arrow					●			
Alignment Tools:								
Flag		●		●		●	●	
Target/Means				●				
Cascading				●				
T/M Tree								
QFD		●		●				

The steps of phase 2 are as follows:[1]

---

[1] The step number refers to the master plan of steps as indicated in Figure 9.1. Other steps will be covered in Chapters 10 and 11.

Chapter 9, Hoshin Phase 2: Management Self-Diagnosis

Step 3a: Select target	Purpose	Tools
3-1 Decide on area to work in	select right target	1. pareto chart of problems

3-2 Set measurement item
    single event
    gradual improvement
    composite
    maintenance
	select appropriate measures and expected rate of improvement	2. target/means matrix improvement items by dept. and section
		1. tree
		2. flags

Step 4: Execution		
4-1 Plan for contingencies	Murphy's Law	PDPC
4-2 Plan step by step execution	will aid efficient success	arrow
Step 5: Monthly Self-Diagnosis		
5-1 Review of tasks undertaken		1. KJ of activities
5-2 Analysis of problems (process improvement)		1. ID of problems
5-3 Problem solving		
5-3-1 Identify the problems	Describe where and when the problem occurs and its extent.	brainstorming, flow chart, check sheet, pareto, etc.
5-3-2 Identify the causes	To identify all possible causes and prime candidates.	check sheet, cause & effect, brainstorming
5-3-3 Take corrective action (short-term)	To try a solution to see if it will help.	modeling, force field analysis PDPC, arrow
5-3-4 Verify action taken	See if solution works	pareto, histogram, control charts
5-3-5 Implement the (long-term) correction	Put the final corrective act in place	PDPC, arrow
5-3-6 Verify the action taken	See if long-term solution works	pareto, histogram, control charts
5-3-7 Standardize	Make sure you hold gains	quality characteristics, substitute quality characteristics, daily control
5-3-8 Congratulate the team	Recognize progress	plaques, beer parties, etc.
5-4 Solution of problems		1. tree target/means for solutions

Part 2 Example

This example is a continuation of the example in the previous chapter and represents the next cycle of activity.

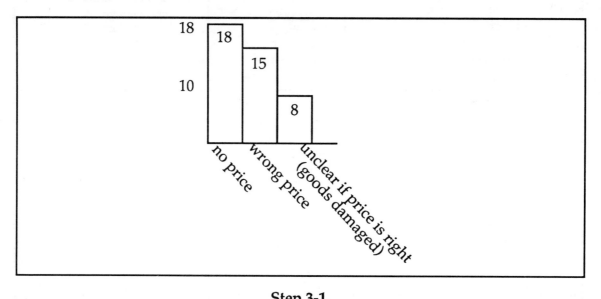

Step 3-1

The most serious problem is the goods that have no price. Although progress was made, the manager is still dissatisified.

Step 3-2-1

'No price' is key. The tree diagram is useful in looking at key causes.

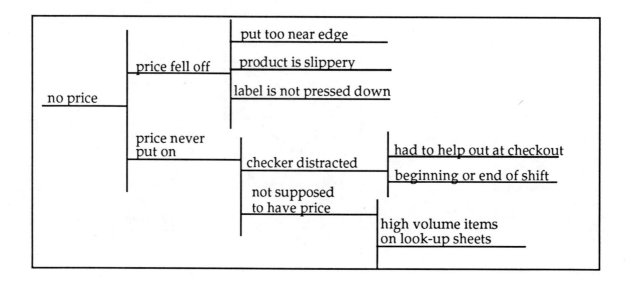

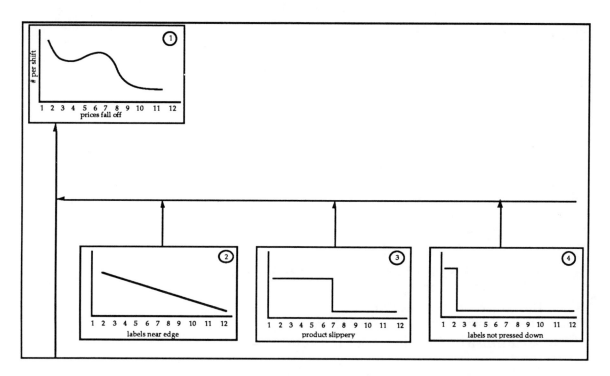

Step 3-2-1

Chart 2 suggests that the problem of labels near the edge will be improved with gradual vigilance, hence a straight line with steady decline.

Chart 3 shows that improvement will be a method or technology change around July. Actually a new label gun will be introduced that works better on frozen goods.

Chart 4 shows a quick method change on how to hold the gun to better press down. This methodology has been known but is not being used by newer employees. A memo with some follow-up training will do the job in February.

Chart 1 is a combination of all three, with a projected decline of 18 to 5 labels that fall off per shift.

Step 4: Execution

4–1-1 Execution - example of PDPC of execution of plan to keep from putting labels near the edge.

Steps in proper label placement
1.0 Research reasons for poor label placement
2.0 Establish standard method
3.0 Develop training
4.0 Train current employees

5.0 Promulgate standards
6.0 Train new employees

Things that can go wrong
1.0 Research reasons for poor label placement
1.1 Research can be incomplete
1.2 Employees can conceal problem
1.3 Important items can be misunderstood
1.4 Items can be misinterpreted by researcher
1.5 Items can be misunderstood by reader

Countermeasures
1.1 Research can be incomplete
1.1.1 Ask all employees what problems they have
1.1.2 Talk to other stores
1.1.3 Talk to trade association
*1.1.4 Talk to vendor of label machine
*1.1.5 Read instructions for label machine

Items 1.1.4 and 1.1.5 look like good countermeasures. Things that can go wrong are identified also for items 2.0 - 6.0 and countermeasures are developed for each of them. This is incorporated into an arrow diagram that may look like this:

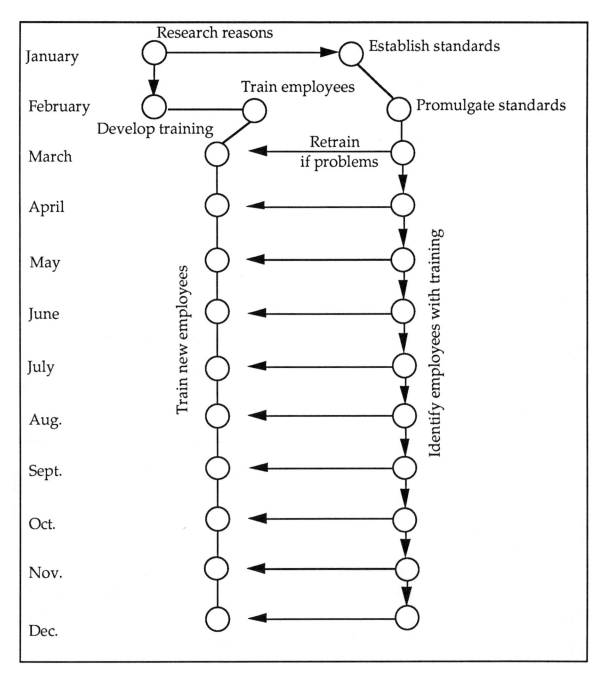

4-2-1 Arrow Diagram

Step 5: Self-Diagnosis

Let's look at one flag for the sake of the example (i.e., labels fall off because they are put near the edge of the package). This is difficult to get a handle on because in some cases it is not clear whether the labels were put on near the edge and fell off or whether the labels were never put on in the first place.

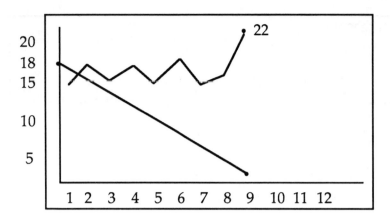

July meeting labels near the top are at 22.

Each manager must monthly diagnose how he or she is doing. This diagnosis should focus on identifying personal and organizational obstacles to planned performance and the development of alternate approaches based on this new information. The following schematic shows the sequence of that activity and the tools that are used.

Monthly Self-Diagnosis

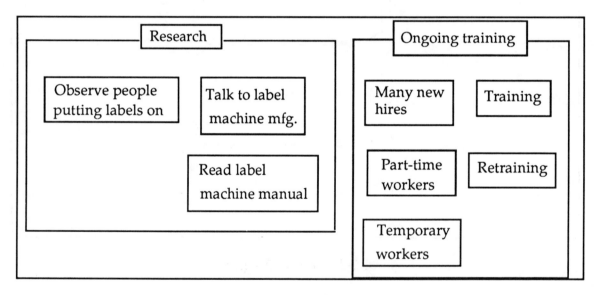

5–1 KJ of Activities

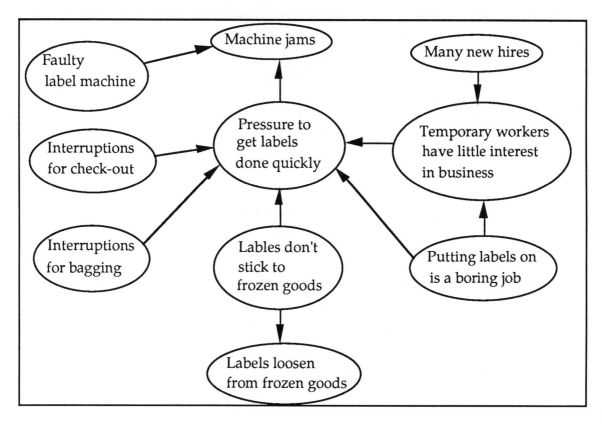

5-2 ID of Problems

Note: The KJ and ID are good tools for bringing things to consciousness and discovering relationships.

5-3 Problem Analysis

Examples of problem analysis are contained in Chapter 8. For the purpose of this discussion let's assume that not allocating proper time for putting labels on and giving the job to people who can't stand it are the major problems.

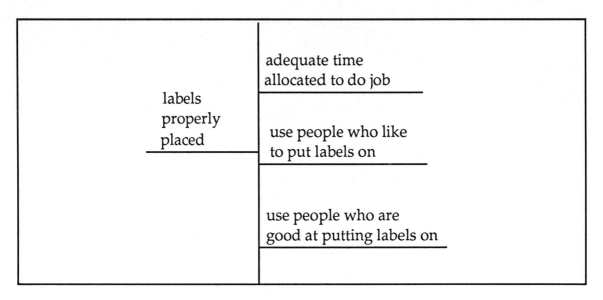

5-4 Tree Target/Means of Solutions

Chapter 10
Hoshin Planning, Phase 3
Alignment of Targets and Means
Throughout the Organization

Objectives make possible direction toward a goal. Process management makes possible the understanding of how variation affects the target. Self-diagnosis makes possible analysis of individual organizational opportunities and obstacles. Alignment, the subject of this chapter, makes possible orientation of various departments in the same direction.

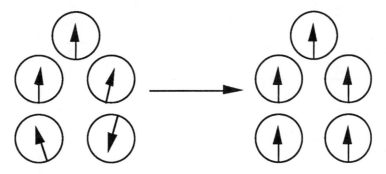

It is not hard to understand how two departments could have different priorities. One level of inventory for example the accounting department may push for minimum inventory and reduced cost, while the manufacturing or operations department may push for maximum inventory and maximum flexibility. When it comes to options on a product or service, the marketing people may want lots of options so as to satisfy more customers whereas manufacturing may want minimal options for ease of manufacturing. It is the purpose of alignment to find the proper balance among such contradictions.

The major steps of Phase 3 are: one year plan, deployment (with alignment), execution, and monthly diagnosis as illustrated in the following chart.

Chapter 10: Hoshin Phase 3 Alignment of Targets and Means

Steps	1	2	3 Deployment individual	3 Deployment align	4 Execution (Process Mgmt)	5 Monthly Diagnosis	6 Annual Diagnosis	Added Dimension Developmental Learnings
	Vision	1 Yr Plan						
(Ch. 8) Phase 1						●		Mgmt by facts
(Ch. 9) Phase 2			●		●	●		Self-diagnosis
(Ch. 10) Phase 3		●	●	●	●	●		Align -3 goals each
(Ch. 11) Phase 4	●	●	●	●	●	●	●	Understand new direction
Tools:	Aim	Plan	Do	Do	Do	Check/Act	Check/Act	
QC:								
Fishbone	●		●			●	●	
Pareto	●				●	●	●	
Line	●				●	●	●	
Flow	●		●			●	●	
Check Sheet	●					●		
Histogram	●					●		
Control	●				●	●		
7M: KJ	●	●						
ID	●	●					●	
Tree	●	●	●				●	
Matrix	●	●					●	
MDA								
PDPC					●			
Arrow					●			
Alignment Tools:								
Flag		●		●		●	●	
Target/Means				●				
Cascading				●				
T/M Tree								
QFD		●			●			

The steps of Phase 3 are outlined as follows:

Chapter 10: Hoshin Phase 3 Alignment of Targets and Means

	Purpose	Tools
Step 2: One Year Plan		
2-1 Analysis of external factors competition, economy	Identify focus based on external issues	QFD ID means tree of means
2-2 Analysis of past problems		KJ of likely tasks ID of target means tree of means
2-3 Zeroing in on key internal issues	Identify focus based on internal problems	KJ of problems matrix by departments ID and tree of means
2-4 Prioritize objectives	Find 2 or 3 priority items	matrix of means vs. quality, cost, delivery
Step 3: Select Target		
3-1 Establish means to reach objectives for self, boss, subordinate	Develop self-initiated plan	target/means tree target/means matrix
3-2 Test plan against objectives and past results	Alignment	matrix review by departments
3-3 Finalize goals and measures	Select appropriate measures	tree diagram flags
3-4 Combine with master plan and records and finalize the target	Put together master plan	flag diagram
Step 4: Execution		
4-1 Plan for contingencies	Murphy's Law	PDPC
4-2 Plan step by step execution	will aid efficient success	arrow
Step 5: Monthly Self-Diagnosis		
5-1 Review of tasks undertaken		1. KJ of activities
5-2 Analysis of problems (process improvement)		1. ID of problems
5-3 Problem solving		
5-3-1 Identify the problems	Describe where and when the problem occurs and its extent	brainstorming, flow chart, check sheet, pareto, etc.
5-3-2 Identify the causes	To identify all possible causes and prime candidates	check sheet, cause & effect, brainstorming
5-3-3 Take corrective action (short-term)	To try a solution to see if it will help	modeling, force field analysis, PDPC, arrow
5-3-4 Verify action taken	See if solution works	pareto, histogram, control charts
5-3-5 Implement the (long-term) correction	Put the final corrective act in place	PDPC, arrow
5-3-6 Verify the action taken	See if long-term solution works	pareto, histogram, control charts
5-3-7 Standardize	Make sure you hold gains	quality characteristics, substitute quality characteristics, daily control
5-3-8 Congratulate the team	Recognize progress	plaques, beer parties, etc.
5-4 Solution of problems		1. tree target/means for solutions

Several of these steps and tools are illustrated elsewhere - some additional examples are given here.

Step 2 (See Chapter 11, p. 14 for examples)
   Objective #1: Reduce lines by having clear price tags on each product

Step 3
3.1 Establish means to reach objectives
Example 3a ID of Means

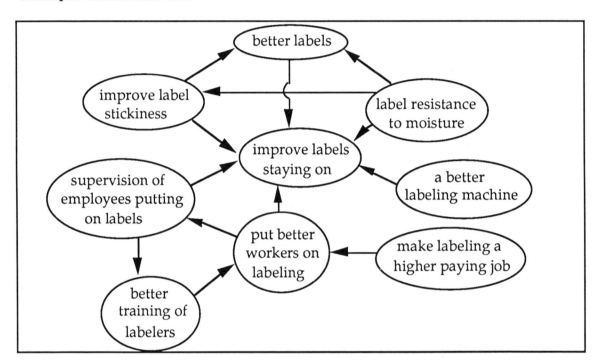

Example 3b

		labels put on all items	labels do not fall off	labels not near edge
	strong relationship ◎			
	some relationship ○			
	possible relationship △			
improve labels staying on	better supervision	◎	△	○
	put better workers on labeling	◎	○	◎
	better labels	△	◎	
	better label machine	○	○	△

3.2 Test plan against objectives and past results
Each department manager shares his matrices in a horizontal quality meeting. During the meeting there is conflict with the purchasing manager who takes issues with improving the labels. He argues that the labels are perfectly alright and that if labelers do their job there will be no problem.

3.3 Finalize goals and measures
   a. Tree diagram

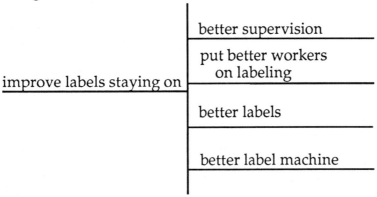

   b. Flags - training of workers

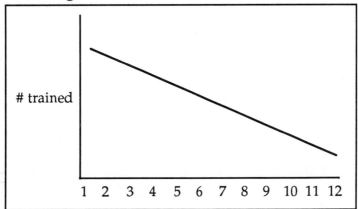

   c. audit of labeling

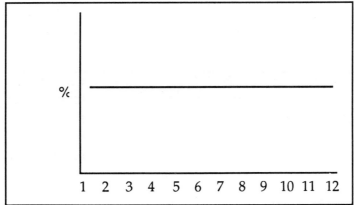

Pareto Chart
Labels that fall off of various goods. Frozen vs. dry goods vs. vegetables

3.4

Once each manager has diagnosed planning results, it is possible to gather results of organizational obstacles and formulate the most critical breakthroughs for the organization from the standpoint of policy or means to accomplish critical targets. This information must be blended with what is happening in the world to develop a strategy for the organization. Some examples of these environmental issues are presented here.

	Major Item		Major Item
(1)	resources, energy (petroleum)	(9)	foods
(2)	foreign currency, yen exchange rate, import restriction, trade friction	(10)	product liability
		(11)	humanism
(3)	pollution, destruction of nature	(12)	international specialization
(4)	change in industrial structure		
(5)	low growth	(13)	technological revolution (software, service, system)
(6)	stronger competition		
(7)	NICS (Newly Industrialized Countries)	(14)	change in age groups
(8)	war power of advanced countries		

Company Environment Changes for Establishment of Management Strategy
TABLE 10.1[1]

	Major Item		Major Item
1)	World-wide market - trade friction, international specialization, local production, internationalization of people, domestic reductions	6)	Use of advanced technology - R&D, electronics, opto-technology, bio-technology

---

[1] Akao, Ibid. Table 2.3.

2) International market - diversification, two-polarization, affluence, highly informed society, change in distribution channel

3) New market development - new product, market size, QCD, vertical and horizontal expansion of current market, technology transfer (use of present technology)

4) Competition - distinction/niche product, production engineering, balance and synchronization, micro fabrication, GT, CAD, CAM, R&D, patent strategy

5) Product liability problems - against users, society

7) Systemization - high value added products, balance between hardware and software, stratification and standardization

8) Higher level schooling and higher age level

9) Pollution - industrial and home life wastes, air pollution, noise, vibration, etc.

10) Others - plant site, energy/resource conservation, international material sourcing

Company Environment Changes for Establishment of Management Strategy

TABLE 10.2[2]

This can lead to a policy statement of where the company is going.

---

[2] Akao, Ibid. Table 2.4.

<Company Motto> Contribute to society with highest quality (Bridgestone) *1
<Management Concept> Achieve customer's satisfaction, everyone's happiness and company's expansion based on quality supreme (Hoyo-Seiki) *2
<Quality Policy> - Thoroughness in customer is No. 1, based on quality-supreme (first clause of management concept, Company) *3 - Understand customers' needs and solve problems. It must be the TI's product and service Let's do correct work from the start. It must be the norm for all of TI's employees. (Nippon T.I. Bipolar Division)
<QC Policy> This company achieves good quality electricity, service, low fee, stable supply, safety and compliance to environment, simultaneously and in the long term so as to respond to customers' needs in a positive manner (Kansai-Electric) *5
<QC Promotion Plan> (omitted)
Notes: Deming Awards 　　　　*1 1968 (Implementation Award) 　　　　*2, *3 1985 (Implementation, Small Industry) 　　　　*4 1985 (Division Award) 　　　　*5 1985 (Implementation Award)

Example of Company Motto, Management Concept, Quality Policy (simple form)
TABLE 10.3[3]

---

[3] Akao, Ibid. Table 2.2.

Chapter 10: Hoshin Phase 3 Alignment of Targets and Means

This is orchestrated throughout the organization using the QC tools for development of targets and means, the alignment tools of the flag system, the matrix of targets and means, and other alignment tools discussed earlier. It is followed by execution plans and a diagnosis of results as described in the previous chapter.

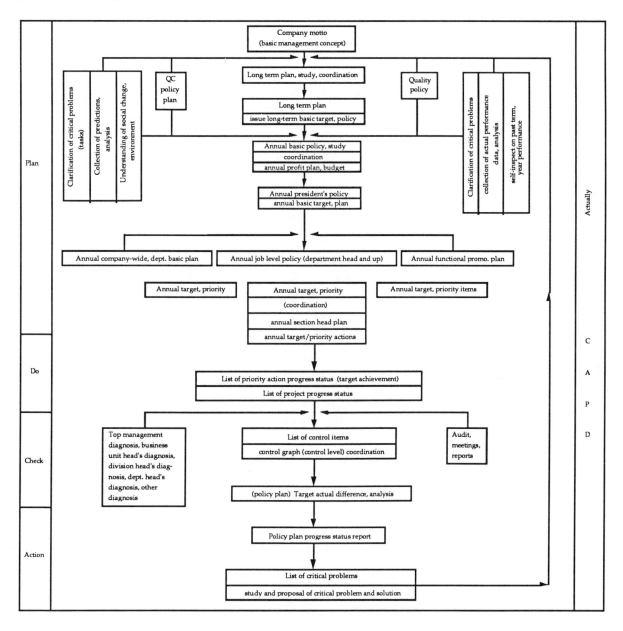

Management by Policy System (basic concept)
FIGURE 10.1[4]

---

[4]Akao, Ibid. Figure 2.3.

Chapter 11
Advanced Hoshin Planning

Advanced hoshin planning uses the seven management tools. This chapter describes each of hoshin planning's six steps in detail and illustrates many of them with a grocery store example.

STEP 1

VISION - 5 Years

Purpose: Develop correct clear vision in a way that will maximize buy-in and results.

sub-steps	purpose	techniques
a. propose company direction	develop a vision	1. KJ of direction
		2. ID of target/means
b. analysis of interference		1. ID of interference factors (managers' last year diagnosis)
set corrective measures		2. Tree of target-means (sent with vision)
c. proposal of vision	test vision	1. Combined tree of target-means (a+b)
d. discussion with department		1. KJ of means by department
review proposal of means		2. Matrix of management vision and department means
e. establish long term vision	finalize vision	

©Bob King, GOAL/QPC

Step 1.a.1 KJ of Direction

Purpose: Many disparate insights and sources of information go into distilling a vision. The purpose of this step is to make sense of those.

Operable question: What do customers expect of us and what do we expect of ourselves long term?

Input: Ideas on vision

Output: Key categories of vision

Tool Used: The tool used is the KJ Chart.

Example explained:
There are many sources of the items included in the chart. Some are from customer surveys, some are from problems we have had in the past, and some are from present opportunities.

Key categories of vision are quality of service, employees are happy, and goods are tracked.

Chapter 11: Advanced Hoshin Planning

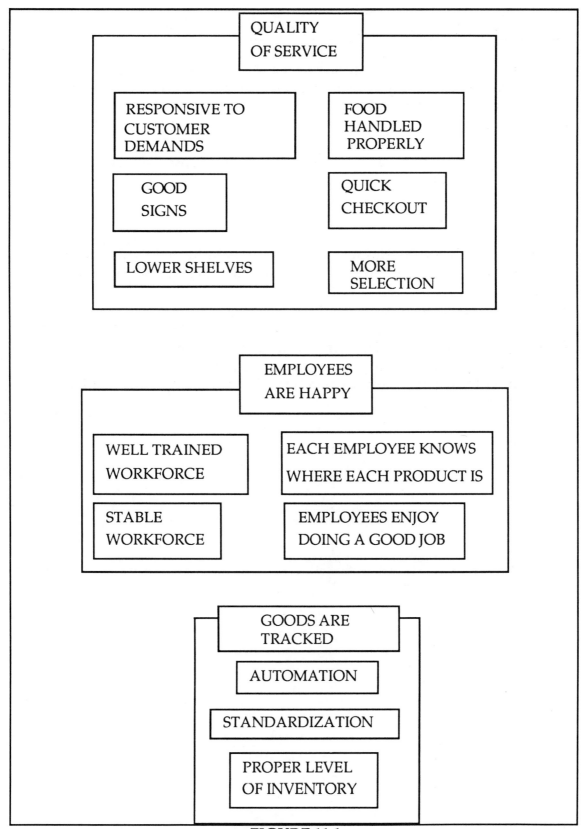

FIGURE 11.1
Step 1.a.1. KJ of Direction

Step 1.a.2 Interrelationship Digraph of Target Means

Purpose: There are many different ways to accomplish a vision. This chart helps identify which items are key because they interact with many other items or because they are initiating items (arrows just going out).

Operable question: What are the means to accomplishing the key categories of the vision?

Input: Means

Output: Key means and initiating means as indicated by numbers of arrows going in and out and just arrows going out respectively.

Tool used: Interrelationship Digraph

Example explained: The two key means to accomplish the vision are "clear procedures" and "customer suggestions". These are also initiating items as are "carts returned to proper location" and "rainchecks for sales."

Chapter 11: Advanced Hoshin Planning p. 11-5

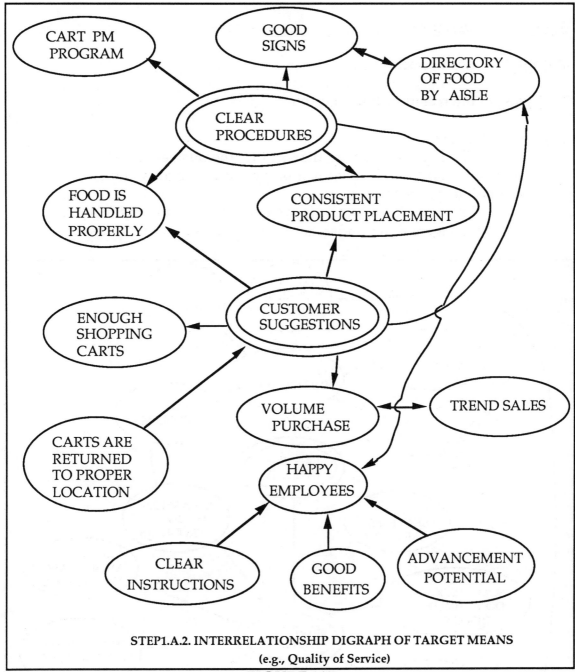

STEP1.A.2. INTERRELATIONSHIP DIGRAPH OF TARGET MEANS
(e.g., Quality of Service)

FIGURE 11.2

Step 1.b.1   ID of Interference Factors

Purpose: It is not enough to dream visions. It is also necessary up front to look at things that can get in the way. This chart identifies what the key obstacles are.

Operable question:
What are the obstacles to accomplishing the key elements of the vision? What are the obstacles to the means to the vision?

Input: Obstacles

Output: Key obstacles

Tool used: Interrelationship Digraph

Example explained:
   The key obstacles are the fact that the business is labor intensive and there is low unemployment in the area and so it is difficult to get good workers.
   The other key obstacle is system problems and the fact that some managers don't understand how to work on the system.

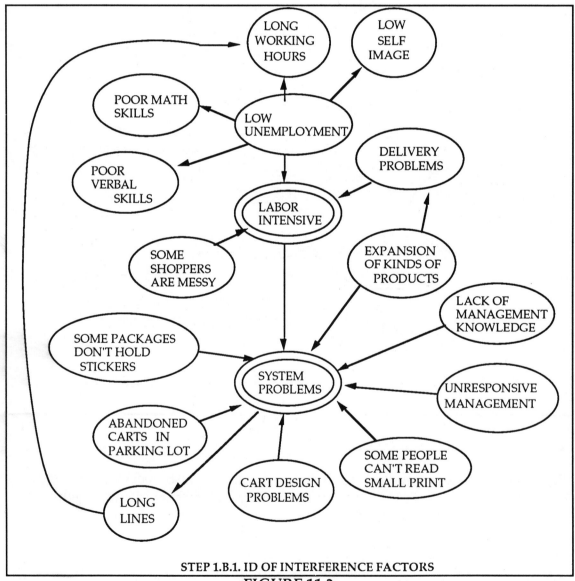

STEP 1.B.1. ID OF INTERFERENCE FACTORS
FIGURE 11.3

Chapter 11: Advanced Hoshin Planning p. 11-7

1.b.2 Tree of Target Means

Purpose: To prepare a systematic picture of the various levels of detailed steps required to overcome obstacles.

Operable question: "How will key obstacles be overcome, how, how, how ?"- leading to a cascading level of detail

Input: Means for improvement

Output: Structured system for improvement

Tool used: System/Tree Chart

Example explained: The key way to deal with system problems is to improve systems.
 The key way to deal with issues related to 'labor intensive' is to improve workers.
 The details of these improvement efforts are portrayed in descending order from general to specific.

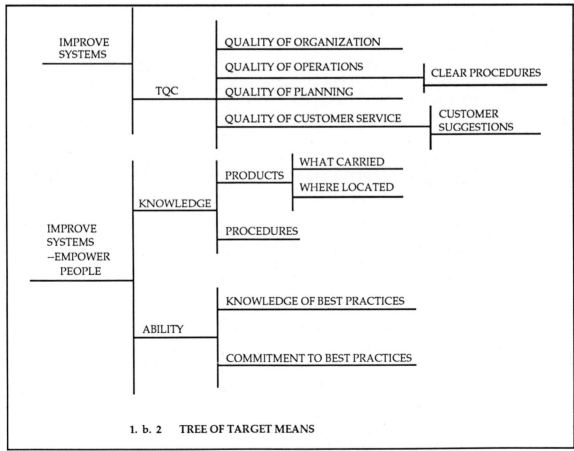

1. b. 2   TREE OF TARGET MEANS

FIGURE 11.4

1-c Proposed Vision

Purpose: To state a draft vision as a basis for more widespread input from managers throughout the organization.

Operable question: Based on analysis of the last four charts, what should our purpose be?

Input: Ideas from previous charts

Output: Draft vision

Tool used: Creative writing

Example explained: The proposed vision details some of the aspects of improved workers and systems.

1 c. PROPOSED VISION

- o ESTABLISH MANAGEMENT TO GROW STEADILY WITH TRU-SAVE AND CONTRIBUTE TO A RICH SOCIETY

- o BE #1 IN QUALITY AND OFFER GOOD PRODUCTS IN RESPONSE TO REQUESTS OF SOCIETY AND TRUST OF CUSTOMERS.

- o IMPROVE MANAGEMENT EFFECTIVENESS BY SYSTEMATIC PROCESS IMPROVEMENT AND CONTROL TO STRENGTHEN COMPANY STRUCTURE.

- o MAKE TRU-SAVE MORE LIKE TRU-LOVE IN MINDS AND HEARTS OF CUSTOMERS.

1. Offer highest quality products and superior services that satisfy customer expectations.

2. Establish an internal environment of mutual trust, development, and growth.

3. Focus on continuous improvement in everything we do.

Chapter 11: Advanced Hoshin Planning   p. 11-9

>Help our customers eat healthier.
>Help our customers eat quicker.

Mission: What company stands for. Society, employees.
Vision: Something we will do within the mission; desired future state.

1.d.1   KJ of Means by Department

Purpose: Develop action plans related to proposed vision and organize them into main categories.

Operable question: What can we do in our area to meet the vision?

Input: Proposed vision

Output: Proposed action plans

Tool used: KJ

Example explained: Details some of the actions to be taken. Note that references to stock people are out of line, because they are not related to what the checkout people can do.

Key activities are improve stock people, study Deming, eliminate obstacles, and improve scheduling.

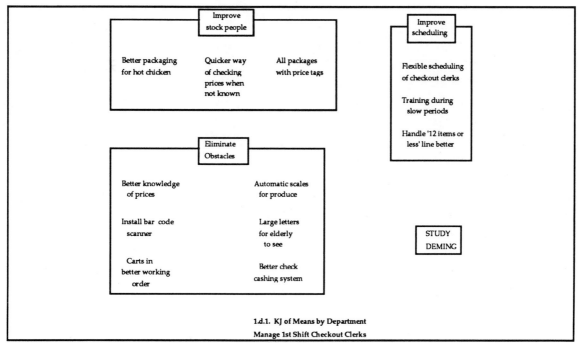

1.d.1. KJ of Means by Department
Manage 1st Shift Checkout Clerks

FIGURE 11.5

1.d.2 Matrix of Vision and Manager Plans

Purpose: To find out how well proposed vision would work if adopted.

Operable Questions: Which implementation plans correlate with which vision statements?

◎ Strong relationship

△ Possible relationship

○ Some relationship

Input: Vision

Output: Analysis of effectiveness of vision

Tool Used: Matrix

Example Explained: All visions seem to be met with quality plans. However, departments are suggesting what others need to do. Focus needs to be more on what initiative each person can take themselves.

Chapter 11: Advanced Hoshin Planning p. 11-11

Matrix of Management Vision and Department Means

◉ Strong relationship

○ Some relationship

△ Possible relationship

	Management Growth	#1 in Quality	Systematic Process Improvement	Improve Image
Study Deming	◉	◉	◉	
Improve Schedule	○			○
Training During Slow Periods	◉			△
Improve Stock People	○			○
All Packages with Price Tags		○	△	○
Automatic Scales for Produce	△	○		○
Install Bar Code Scanner	○	◉	◉	◉
Better Check Cashing System	○		○	

FIGURE 11.7

Step 2: Set 1 Year President's Plan

1. Review previous year
2. Environmental analysis
3. Review long range plan
4. Integrate above three items

STEP 2:

1 Year Plan

Purpose: To state three or four key items for the year in a way that will maximize success

sub-steps	techniques
a. clarify annual task based on vision	1. ID of means 2. Tree of means

b. analysis of factors internal and external to the company
   e.g., economic statistics
      forecasted social change
      engineering info. on competition
      engineering info. patents
	1. clarification of expected tasks by KJ
	2. ID of target-means
	3. Tree of target means

c. setting of tasks to be solved c-1 analysis of company-wide problems by president's annual audit	1. KJ of problem functions 2. task/function matrix per dept.
c-2 understand plan vs. actual of whole company and depts. analyze gaps by ID of problem factors	1. flag results for year
c-3 deploy each functional task	1. ID 2. Tree
d. arrange/conform above	1. matrix of a,b,c, means vs. quality, cost, delivery problem of conflicts

2.a.1 ID of means to vision to help establish one year plan.

Purpose: Identify key projects for year based on vision

Operable Question: What are the key actions that must be taken (based on vision)?

Input: Action items

Output: Prioritized action items

Tool used: Interrelationship Digraph

Chapter 11: Advanced Hoshin Planning p. 11-13

Example explained: Two key activities are educating all employees and updating standards because they have the most arrows. (8 and 5 respectively)

Use catch ball as a key starting action.

2.a.1. ID of Means to vision to help establish one year plan
- educate all employees
- visit other successful stores
- learn from competitors
- learn from customers
- get customer feedback
- educate customers
- educate suppliers
- establish standards
- update standards
- practice continuous improvement
- use better record keeping
- use better planning
- empower employees
- use catch ball

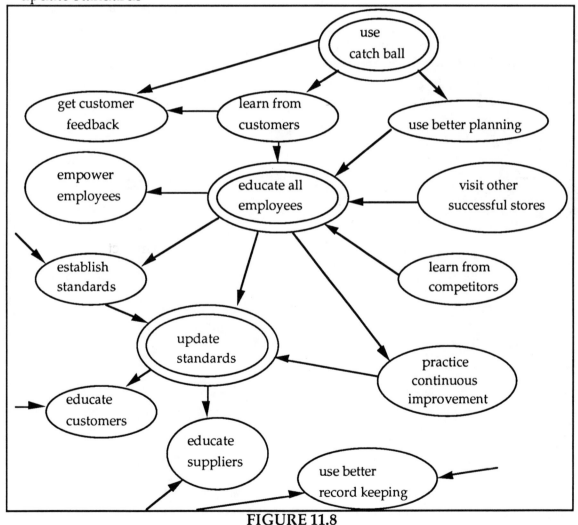

FIGURE 11.8

2.a.2 Tree of Means to Help Develop One Year Plan

Purpose: Detail action steps

Operable Question: How is each action step accomplished? Build tree going from general to specific

Input: Action items from previous chart

Output: Detailed plan

Tool Used: Tree Diagram

Example Explained: Survey forms, focus groups. Comparison shopping etc., are some of the key individual steps in educating all employees.

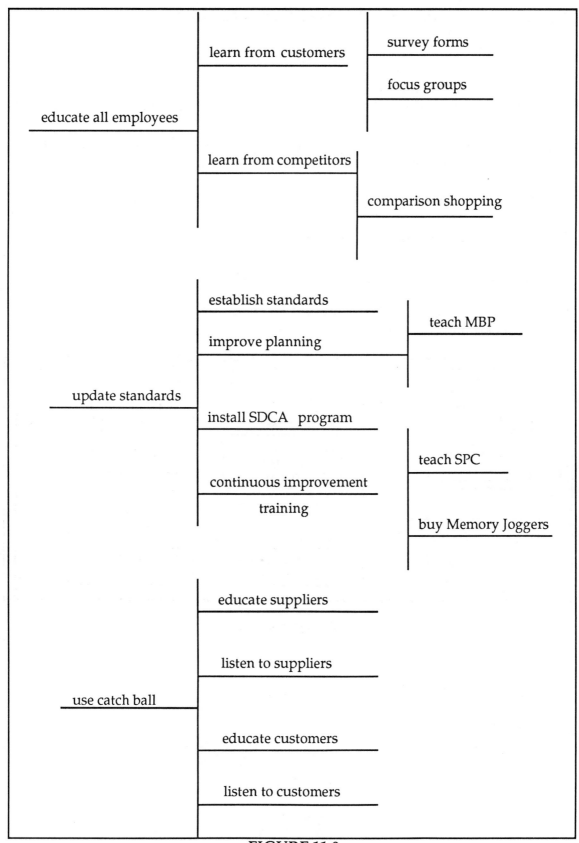

FIGURE 11.9

2.b.1 Clarification of Expected Tasks by KJ

Purpose: To examine internal and external environment and bring order to potential tasks.

Operable Question: What tasks are required by what is going on inside and outside company? (environment question)

Input: Expected tasks

Output: Key categories of expected tasks

Tool Used: KJ

Example Explained: Key required tasks are improved recruiting, checkout and cart system

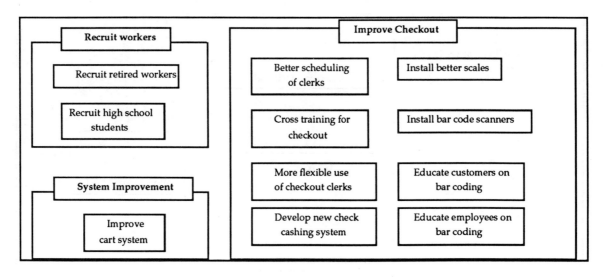

FIGURE 11.10

Chapter 11: Advanced Hoshin Planning p. 11-17

2.b.2 ID of Target Means

Purpose: Discover best means to accomplish tasks identified in previous chart.

Operable Question: What are the means to accomplish these tasks?

Input: Means to accomplish tasks

Output: Priority of means

Tool Used: Interrelationship Digraph

Example Explained: One of the means for improved checkout (i.e., bar coding) is explained. The right equipment and the right training emerge from the ID as key items to work on.

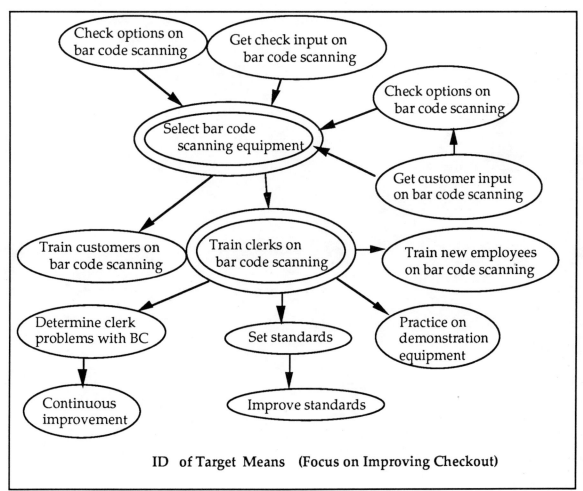

ID of Target Means (Focus on Improving Checkout)

FIGURE 11.11

2.b.3 Tree of Target Means

Purpose: To delineate the system for accomplishing the key tasks.

Operable Question: 'How accomplished?' is asked as the system is laid out going from general to specific.

Input: Key tasks.

Output: System to accomplish tasks.

Tool Used: Tree Diagram

Example Explained: The Tree Diagram takes the key items of selecting bar coding and training checkout clerks on bar coding and systematically lays out how they will be accomplished using a Tree Diagram.

Chapter 11: Advanced Hoshin Planning

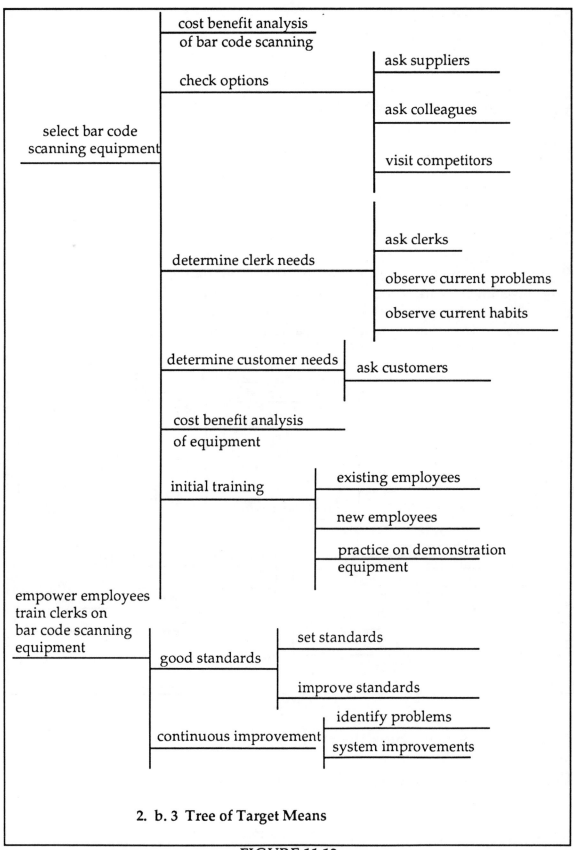

2. b. 3 Tree of Target Means

FIGURE 11.12

2.c.1 KJ of Problem Functions

Purpose: To examine last year's problems

Operable Question: What company problems kept managers from succeeding last year?

Input: Problems

Output: Major categories of problems.

Tool Used: KJ

Example Explained: In reviewing the previous year, management weaknesses are listed and organized using the KJ Method. Those that emerge as key are new management methods, listening better to customers, handling changing business situations, etc.

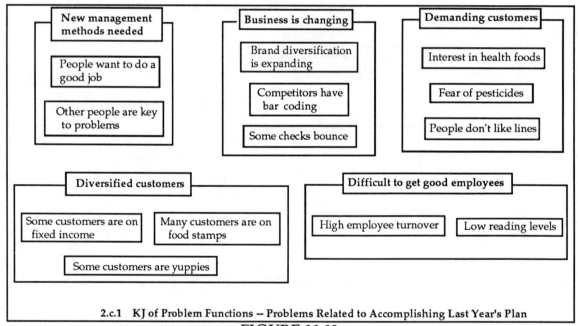

2.c.1 KJ of Problem Functions -- Problems Related to Accomplishing Last Year's Plan
FIGURE 11.13

2.c.2 Task Function Matrix Per Department

Purpose: To identify where the major management problems existed.

Operable Question: What is the relationship between each problem and each department?

Chapter 11: Advanced Hoshin Planning p. 11-21

Input: Problems and departments.

Output: Which departments had what problems?

Tool Used: Matrix

Example Explained: Management challenges are reviewed by department using a Matrix Chart.

2.c.2 Task Function Matrix per Department
Based on 2.c.1

Management Function Problems

	Courteous to Customers	Knowledgeable about Products	Helpful to Customers	Help Employees Learn	Help Employees Grow	Safety Better Problem Solving Skills	More Product Knowledge
Ordering Dept.							
Stock Room							
Pricing Dept.							
Check Out							
Produce							
Frozen Goods							
Dry Goods							

FIGURE 11.14

2.c.3 Flag Results for Year

Purpose: To examine specific results from previous year as spelled out in Flag Charts.

Operable Question: Why did we meet targets or fail to meet targets?

Input: Flag Chart

Output: Analysis of Flag Chart.

Tool Used: Flag system, line graphs.

Example Explained: Specific projects of the years are analyzed. This example is that of labels falling off. Half-way through the year it was recognized that bad adhesive on the labels caused them to fall off, particularly on frozen foods. Purchasing had bought from the lowest bidder. The important issue is how can problem solving be improved for next year? What system problems led to buying from the lowest bidder?

Chapter 11: Advanced Hoshin Planning p. 11-23

2.c.3 Flag Results for Year

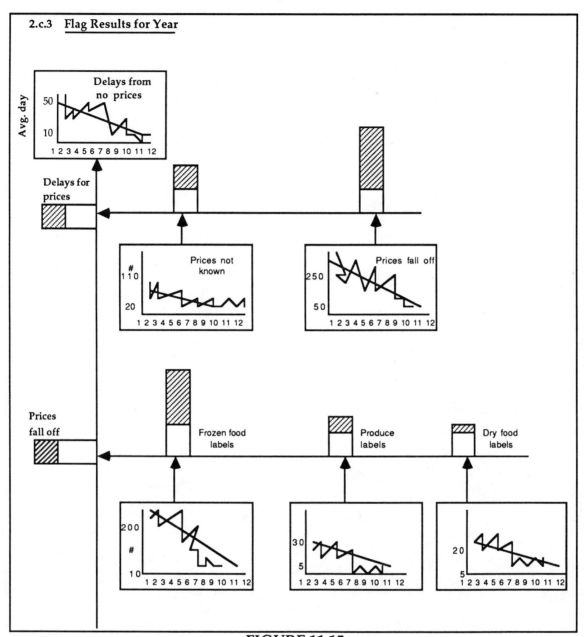

FIGURE 11.15

2.c.4. ID of Problem Factors

Purpose: Analyze past year's problem to discover key problems.

Operable Question: What are the reasons why we got into trouble or reasons why we didn't do better?

Input: Causes of shortcomings.

Output: Prioritization of causes of shortcomings.

Tool Used: ID

Example Explained: This ID Chart shows that lack of planning was one of the key factors that led to buying from the lowest bidder.

✓ 1. Lack of planning

✓ 2. Lack of data gathering

✓ 3. Failure to use flow charts

✓ 4. Heavy handed supervision of clerks

✓ 5. No time for education

6. Difficult for clerks to remember all prices

7. Price changes due to sales

8. Cannot use price tags on vegetables

✓ 9. Adversarial relationship with suppliers

✓ 10. Buying supplies based on cost only

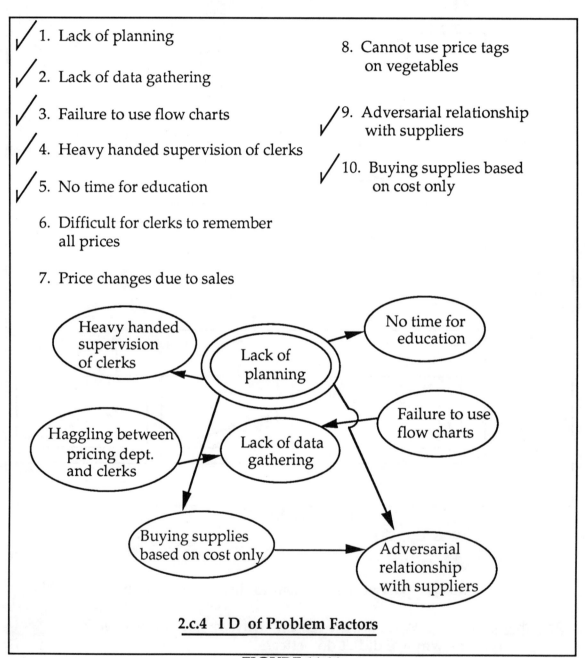

2.c.4   I D of Problem Factors

FIGURE 11.16

2.c.5 Tree Diagram to Deploy each Functional Task

Purpose: To develop a plan to deal with causes of short comings.

Operable Question: What are key plans? What are the systems to accomplish them?

Input: Action plans.

Output: Organized action plans.

Tool Used: System/Tree Chart

Example Explained: This chart shows how two improvement areas are spelled out using the Tree Diagram. The examples used are improved planning and improved pricing system. In the area of improved pricing system, the bar coding emerges as key.

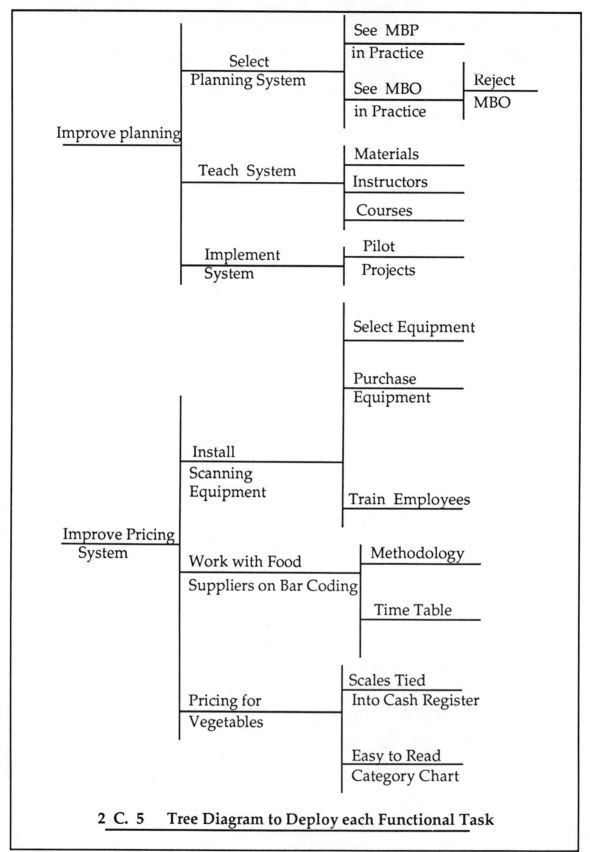

2 C. 5   Tree Diagram to Deploy each Functional Task

FIGURE 11.17

Chapter 11: Advanced Hoshin Planning   p. 11-27

2.d.1   Matrix of ABC Means vs. Quality Cost Delivery

Purpose: Bring together potential plans from three areas: vision analysis, environment analysis, last year's analysis. Identify potential measures for key analysis action items.

Operable Question: Which proposed actions relate to quality, cost, delivery, and education?

Input: Action items.

Output: Grouping of action items.

Tool Used: Matrix

Example Explained: Several of the plans are listed and analyzed to see how they relate to the major category measures of quality, cost, delivery, and education.

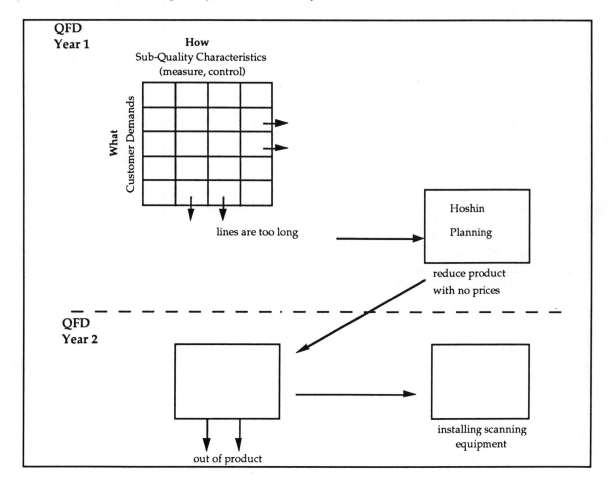

FIGURE 11.18

Step 2:   Goals for Year

Purpose: State one year goals.

Operable Question: Based on analysis of vision, environment and last year, what are our goals for next year?

Input: Preceding analyses.

Output: One year goals.

Tool Used: Analyses and creative writing.

Example Explained: The goals for the year are listed.

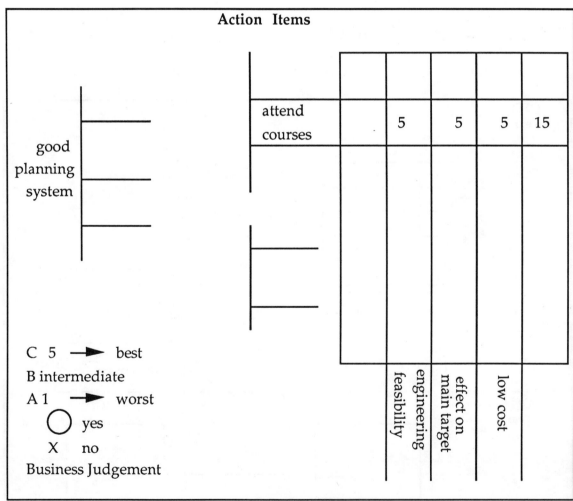

FIGURE 11.19

Chapter 11: Advanced Hoshin Planning

```
                                    Quality  Cost  Delivery  Education
  1 Year Plans

  Courteous to Customers

  Knowledgeable about
       Products

  Helps Employees Grow

* Improve Planning             ◎     ◎     ◎      ◎

  Improve Pricing System
```

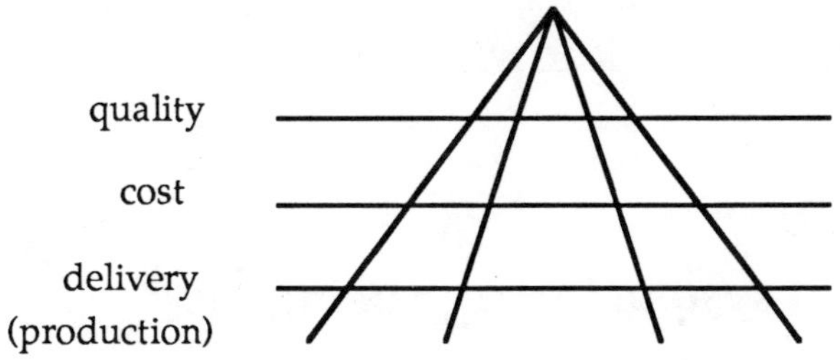

quality
cost
delivery
(production)

2.d.1 Matrix of ABC Means vs. Quality Cost Delivery

FIGURE 11.20

Step 2:   Goals for Year

Purpose: State one year goals.

Operable Question: Based on analysis of vision, environment, and last year, what are our goals for next year?

Input: Preceding analyses.

Output: One year goals.

Tool Used: Analyses and creative writing.

Example Explained: The goals for the year are listed.

©Bob King, GOAL/QPC

STEP 2

Conclusion

OBJECTIVES FOR YEAR

1. Improve checkout through installation of bar code scanning equipment and new scales for produce.

2. Improve customer service through more customer input and more extensive product knowledge.

3. Improve management system through using the M B P approach to planning.

Step 3:   Deployment to each department

    1.   Review previous year's process

    2.   Target/measure chart

    3.   Daily control plan

Chapter 11: Advanced Hoshin Planning p. 11-31

STEP 3

DEPLOYMENT TO DEPARTMENTS

<u>Purpose:</u> Assure that the plan is operationalized in each department with the subordinate having maximum responsibility for setting the plan and making it work.

<u>sub-steps</u>   <u>techniques</u>

a. establish means to reach objectives   1. objective/measures tree
    for manager, supervisor, employee   (end/means)

                                                2. target/measures matrix

b. review components of item to be   1. pareto chart of improve-
    improved by department   ment items by dept. and
                                                    section

    b.1   based on annual plan
    b.2   based on previous results

c. set goals by department and section   1. Tree diagram
    with monthly measures
    - single event   2. Flags
    - gradual improvement
    - composite
    - maintenance

d. combine with master plan and   1. Flag diagram
    master record sheets

STEP 3

3.a.1   Tree Diagram of means for department to reach one year goals.

<u>Target Measures Matrix to Establish Appropriate Measures</u>

<u>Purpose:</u>   To select proper measures.

Operable Question: Which measures correlate with which action steps?

Input: Steps and measures

Output: Prioritization of measures.

Tool Used: Matrix

3.a.2   Target measures matrix to establish appropriate measures

Matrix (pareto) Analysis of Value of each Proposed Target

Purpose: Prioritize and select appropriate tasks.

Operable Question:
1) Which action steps will contribute to desired effect?
2) engineering feasibility of action steps
3) return on investment of action steps
4) which action steps best meet these three criteria?

Input: Action steps and criteria

Output: Prioritized action steps

Tool Used: Tree and Matrix

3.b.1   Matrix (pareto) analysis of value of each proposed target.

Tree Diagram of Goals

Purpose: Add goals to Tree Diagram

Operable Question: What are appropriate targets for each action item?

Input: Action items and measures.

Output: Targets for action items

Tool Used: Tree Diagram

Chapter 11: Advanced Hoshin Planning p. 11-33

3.c.1　　　　Tree Diagram of goals, target and measure

<u>Establish Flags for Goals</u>

<u>Purpose:</u>　　Set goals and slope of improvement.

<u>Operable Question:</u>　　What is realistic goal, will it be accomplished by continuous improvement, innovation, or standardization?

<u>Input:</u>　　Action items

<u>Output:</u>　　Monthly targets

<u>Tool Used:</u>　Line Charts

3.c.2　　　　Establish flags for goals

<u>Flag Diagram</u>

<u>Purpose:</u>　　Coordinate activities

<u>Operable Question:</u>　　How are flags arranged systematically? Are there any inconsistencies or conflicts?

<u>Input:</u>　　Flags

<u>Output:</u>　　Flag system

<u>Tool Used:</u>　Flag system

3.d.1　　　　Flag Diagram

Step 4:　　　Execution of Plan

　　1.　Each person down through quality circle helps president and immediate supervisor to meet their goals

STEP 4:

EXECUTION

Purpose: To execute in a way to assure success at all levels

Sub-Steps   Techniques

a. plan detailed execution   1. Arrow Diagram
    based on established routines

b. do contingency planning   2. PDPC (Process Decision
                                                   Program Chart)

c. construct execution report   3. Revised arrow diagram

Chapter 11: Advanced Hoshin Planning

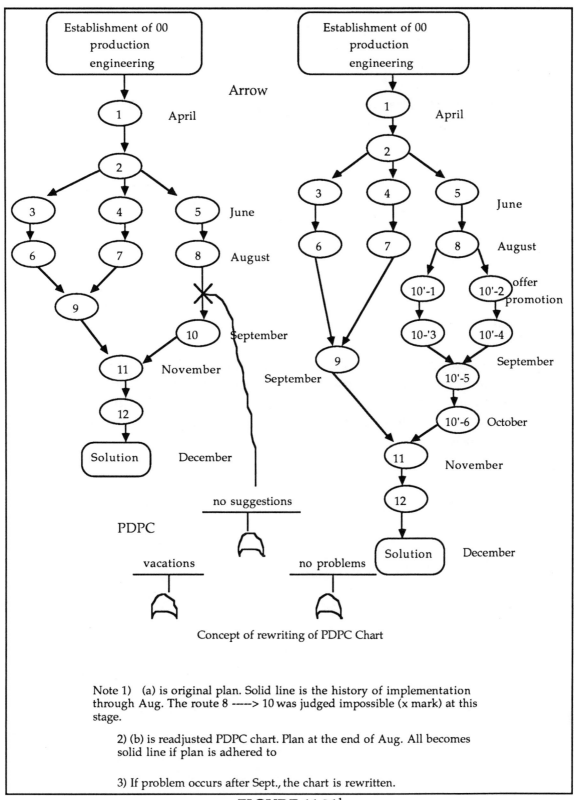

Note 1) (a) is original plan. Solid line is the history of implementation through Aug. The route 8 -----> 10 was judged impossible (x mark) at this stage.

2) (b) is readjusted PDPC chart. Plan at the end of Aug. All becomes solid line if plan is adhered to

3) If problem occurs after Sept., the chart is rewritten.

FIGURE 11.21[1]

---

[1]Nyatani, Ibid. Figure 3.3.

PDPC of Plan

Purpose: - Set up required steps

- Analyze things that can go wrong and develop countermeasures.

Operable Questions: 1) What are key steps in carrying out activity?
2) What can go wrong at each step?
3) What are potential countermeasures?
4) What are optimal countermeasures?

Input: Experience

Output: Plan

Tool Used: PDPC

Example Explained: The example is on the installation of bar code scanning equipment. Problems that can occur at the step of training employees are listed. Employee apathy is examined. Countermeasures are listed. Focus on big picture and better management are selected to counteract apathy.

PDPC

STEPS
1. IDENTIFY NEEDS (CUSTOMER)
2. SELECT EQUIPMENT
3. TRAIN EMPLOYEES
4. INSTALL EQUIPMENT
5. HAPPY CUSTOMERS

PROBLEMS
- 3.0 TRAINING
- 3.1 NO DEMONSTRATION EQUIPMENT
- 3.2 EMPLOYEE APATHY
- 3.3 NOT ENOUGH TIME
- 3.4 NO MANUALS
- 3.5 TRAINING TOO LATE
- 3.6 TRAINING TOO EARLY

COUNTERMEASURES
- 3.2.0 EMPLOYEE APATHY
- 3.2.1 MORE MONEY
- 3.2.2 NO LAYOFFS
- 3.2.3 FIRE THEM
- *3.2.4 BETTER MANAGEMENT
- **3.2.5 BIG PICTURE

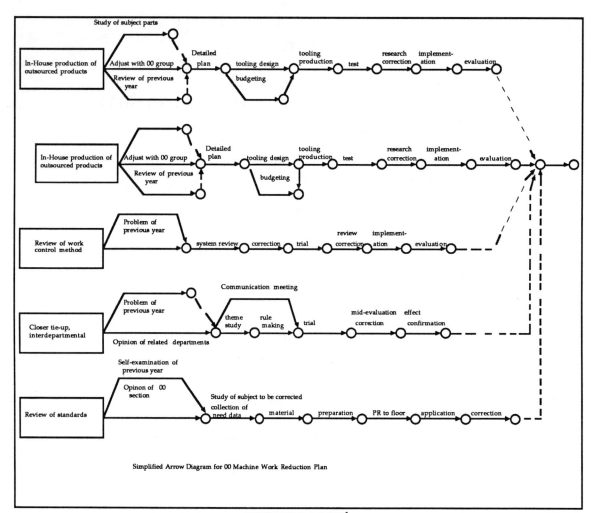

FIGURE 11.22[1]

Step 5:   Monthly Audit

1. Review control graphs for
      quality
      cost
      delivery

2. Focus on process rather than targets

STEP 5:
AUDIT

Purpose: To improve the planning process

---

[1]Nyatani, Ibid. Figure 3.32.

Look at results and analyze process

Sub-Steps

1. Review management functions of each department
2. Arrangement of tasks
3. Analysis of problems
4. Solution of problems

Techniques

1. Dept./Function Matrix
2. KJ
3. ID of problems
4. Target/means system chart of solutions

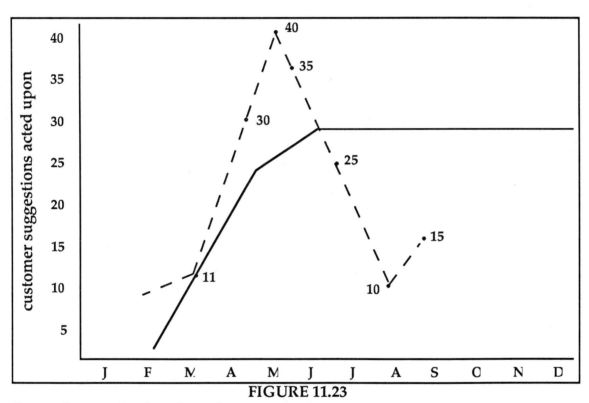

FIGURE 11.23

Jan. Suggestion box is under construction.
Forms have been developed and are at the printer.

Feb. Suggestion box installed on February 5th. Twenty suggestions are submitted. Most are related to requests for specialized products, five for new wines, two for Poulley-fuisse, two for Mouton-Cadet, and one for Chateauneuf du Pap. Six suggestions are related to special product. Four related to new cuts of meat. Five related to better signs.

Suggestions implemented:

a. Four cuts of meat - sign put up for cuts of meat
b. Four special produce ordered and delivered

Suggestions not implemented

a. Two special produce not available in February (2)
b. Better signs will be studied (5)
c. New wine takes one month lead time (5)

Summary: 20 suggestions received - eight implemented

Chapter 11: Advanced Hoshin Planning

Mar. 40 suggestions received, 11 implemented

	From last month:	Action Taken	Action Postponed	Action Not Possible
carry over suggestions	(2) special produce (5) better signs (5) new wine	(2) special produce stocked (4) wine ordered	(5) understudy	Chateau neuf-du-pape not available from producer
new suggestions	(15) for better signs (15) for shorter check-out lines (5) better service in weighing vegetables (3) better bagging (2) not to many heavy things in each bag	(3) baggers told to be more systematic (2) baggers told to use more bags	(15) better signs under study (15) shorter check out lines under study (5) weighing program under study	
Total	52	11	40	1

Assessment: Many items being postponed. Bagging problem may be a quick fix rather than a system improvement. Line waiting should be improved with bar coding reading equipment.

Apr.

From last month:	Action taken	Action postponed	Action not possible
(20) better signs (15) shorter check out lines (5) improve weighing	(20) new signs installed	(15) bar coding coming (5) weighing understudy	
(10) better new signs (20) shorter check out lines (10) too many bags (20) weighing problems	(10) new sign installed	(20) bar coding coming (10) investigate bagging (20) investigate bagging	
100	30	70	File: customer, pg 5

FIGURE 11.24a

May

Action from last month	Action taken	Action postponed	No action
(35) better check out lanes (10) too many bags (25) weighing delays	(35) bar coding readers ordered and received	(10) bagging under study (25) weighing under study	
New suggestion (5) too much waiting at check out (15) too many goods out of stock (3) goods too high on shelf for short people	(5) bar coding readers ordered and received	(15) ordering process understudy	(3) no space but up
93	40	50	3

Assessment: Action taken ordering the bar coding readers. Other alternative have been found to arrange goods for short people. Example, narrower columns with extra stand up top.

June

Action from last month	Action taken	Action postponed	No action possible
(10) too many bags (25) weighing delays (15) goods out of stock		(10) bagging under study (25) weighing under study (15) goods out of stock	
New (35) prices not on shelves problem related to bar coding	(35) pricing person told to do a better job		

Assessment: Many bar coding problems are taking energy and time from other projects.

File: customer,pg6

FIGURE 11.24b

Chapter 11: Advanced Hoshin Planning   p. 11-43

July

Action postponed	Action taken	Action postponed	No action possible
(10) bagging problems (15) goods out of stock		(10) bagging problems being studies (15) goods out of stock studies	
New (25) complaints about bar coding	(25) check out clerks told to do a better job		
50	25	25	

Aug.

Actions postponed	Action taken	Action postponed	
(10) bagging problems (15) goods out of stock	(10) baggers told to do a better job	(15) goods out of stock to be studied	

Sept.

Actions postponed	Action taken	Action postponed	No action possible
(15) goods out of stock	(15) order processing told to study bar coding reading results		
15	15		

Assessment: No suggestions for two months -

Not working on system.
Bar coding problems keeping managers from worrying about other customer complaints.

File: customer,pg7

FIGURE 11.24c

Step 6:   Presidential Audit

    1. Review focusing on process

        to get better plans

        and results

        for following year

STEP 6:

PRESIDENT'S ANNUAL REVIEW

Purpose: Select critical items for next year

Sub-steps	Techniques
1. review successes and failures in planning and execution and audits	1. matrix report of flag results for year
	2. Company, plant, dept., section reports of monthly meetings work on improving the planning process
	1. analysis of problems by I.D.

Getting Started

Questions for Evaluation of Hoshin Planning Implementation

1.   Self-examination of previous year with data

2. Reason for introduction clear

3. Process oriented?

4. Adjustment with other departments? Is adjustment with other departments sufficient?

5. Is target clear and qualified?

6. Is catch-ball, top-bottom sufficient?

7. Is deployment to peripherals sufficient?

8. Is time schedule clear?

9. Is delivery date kept?

10. Is interdepartmental tie-up for implementation sufficient?

11. Is relationship with committees or conferences sufficient?

12. Is it clear who is in charge of each process?

13. Is arrangement for progress check there?

14. Is analysis of difference and target proper and correct?

15. Is evaluation yardstick clear?

16. Were needed presentation or explanation meetings held?

17. Was hearing of upper people held?

18. Was self-examination and summary of results done?

19. Was self-examination for next year done?

Issues to Consider in Implementing Hoshin Planning

1. Skill of company

2. The will of top management

3. Leadership of top management

4. Attitudes of top management toward TQC

5. Attitudes of top management toward hoshin planning system

6. The will of the promotion department

7. Capability of the promotion department

8. Will of implementation departments

9. Capability of implementation departments

Implementation guidelines of hoshin planning

No TQC

1. Top management plan

2. Set improvement tasks for each department

3. Solve problems using Seven QC Tools and Seven Management Tools. As people get experience with these tools they will be more comfortable with them.

4. After one or two years people can go into formal process described in six steps.

In case where there has been QC education

Prerequisites in place
1. QC circles in place
2. Seven QC tools in place
3. SQC education is being done
4. Level of specific technologies is high
5. TQC is being implemented. Plan to get Deming Prize in three to four years.

Method of proceeding

1. Introduce deployment or execution at two or three plants or branches
2. Evaluate whether they are ready for interdepartmental activities (This leads to high morale of employees and good results)
3. TQC implemented in each department to build confidence
4. TQC implemented in the entire company, including corporate head office in general.

Sales and service departments

1. Steps three and four/seven management tools
2. Add numerical data in years three and four
3. Use flag, target/means matrix/seven management tools
4. Use full six steps. Every year hoshin planning should be upgraded but it can only be as good as TQC.

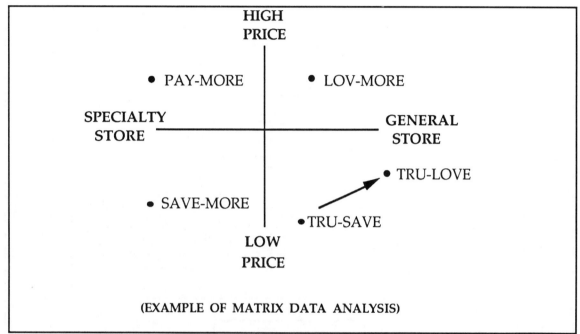

(EXAMPLE OF MATRIX DATA ANALYSIS)

FIGURE 11.25

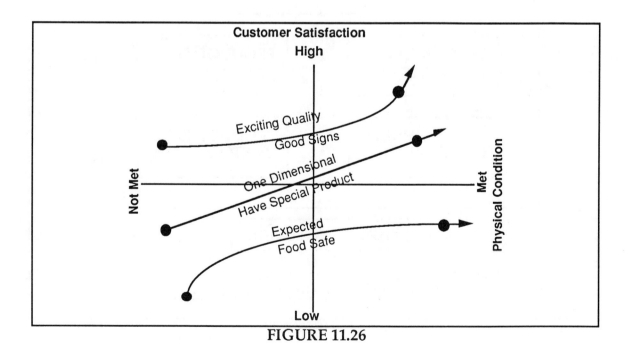

FIGURE 11.26

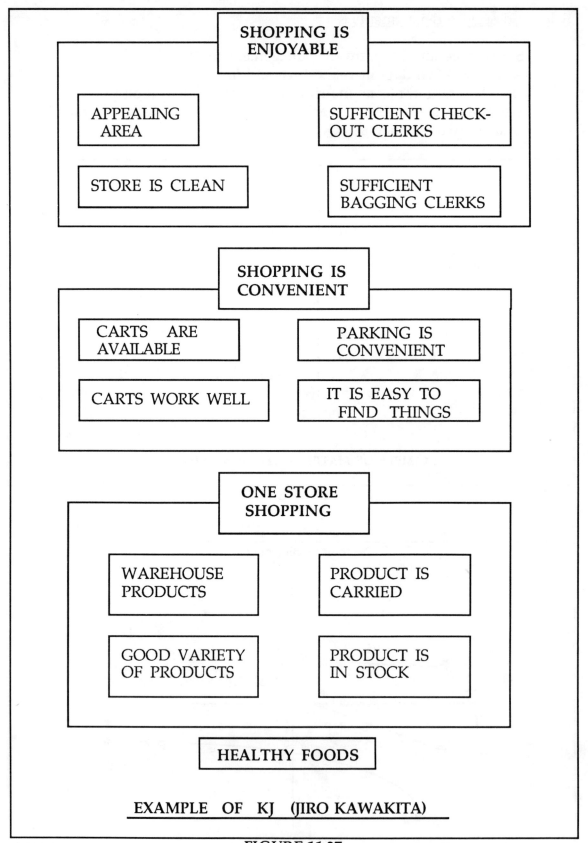

FIGURE 11.27

Chapter 11: Advanced Hoshin Planning

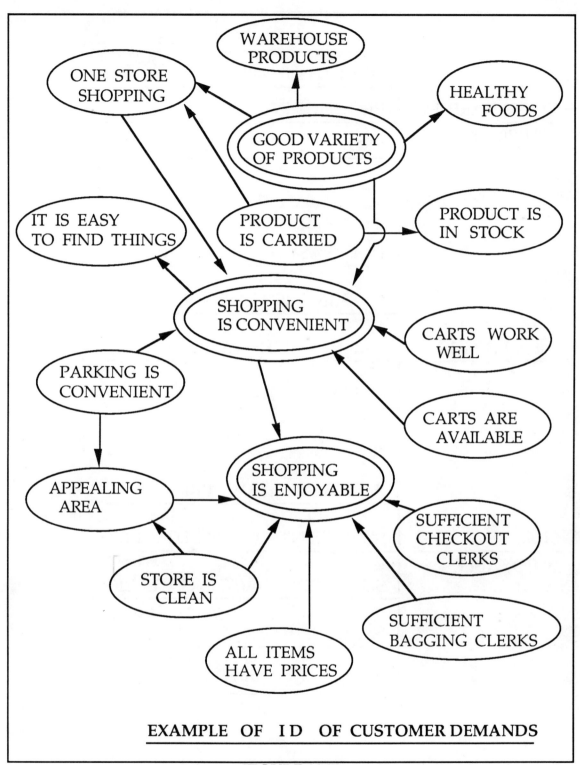

EXAMPLE OF ID OF CUSTOMER DEMANDS

FIGURE 11.28

	WAREHOUSE PRODUCTS
	HEALTHY FOODS
GOOD VARIETY OF PRODUCTS	PRODUCT IS CARRIED
	PRODUCT IS IN STOCK
	IT IS EASY TO FIND THINGS
	PARKING IS CONVENIENT
SHOPPING IS CONVENIENT	CARTS WORK WELL
	CARTS ARE AVAILABLE
	ALL ITEMS HAVE PRICES
SHOPPING IS ENJOYABLE	SUFFICIENT CHECKOUT CLERKS
	SUFFICIENT BAGGING CLERKS
	STORE IS CLEAN
	APPEALING AREA

TREE DIAGRAM OF CUSTOMER DEMANDS

FIGURE 11.29

Chapter 11: Advanced Hoshin Planning

					CUSTOMER DEMANDS				
PLAN	COMPETITOR Y	COMPETITOR X	COMPANY NOW	%	TOTAL	HEALTHY FOODS	ALL ITEMS HAVE PRICES	PRODUCT IS IN STOCK	EASY TO FIND THINGS	
5	4	3	4	30	360				◎ 360	GOOD SIGNS
4	4	3	4	22	270			◎ 270		TREND OF SALES
5	4	4	4	32	390	O 45	◎ 135	O 90	O 120	GOOD PROCEDURES
5	5	3	5	16	210	◎ 135	△ 15	△ 30	△ 40	FOOD HANDLED PROPERLY
						4	3	5	5	RATE OF IMPORTANCE
						5	3	4	4	COMPANY NOW
						3	4	3	3	COMPETITOR X
						5	4	5	4	COMPETITOR Y
						5	4	5	5	PLAN
						1	1.33	1.25	1.25	RATIO OF IMPROVEMENT
								O	◎	SALES POINT
						4	4	7.5	9.4	ABSOLUTE WT.
							15	30	40	DEMANDED WT%

◎ = 9
O = 3
△ = 1

SUBSTITUTE QUALITY CHARACTERISTICS

FIGURE 11.30

Appendix A p. A-1

Appendix A

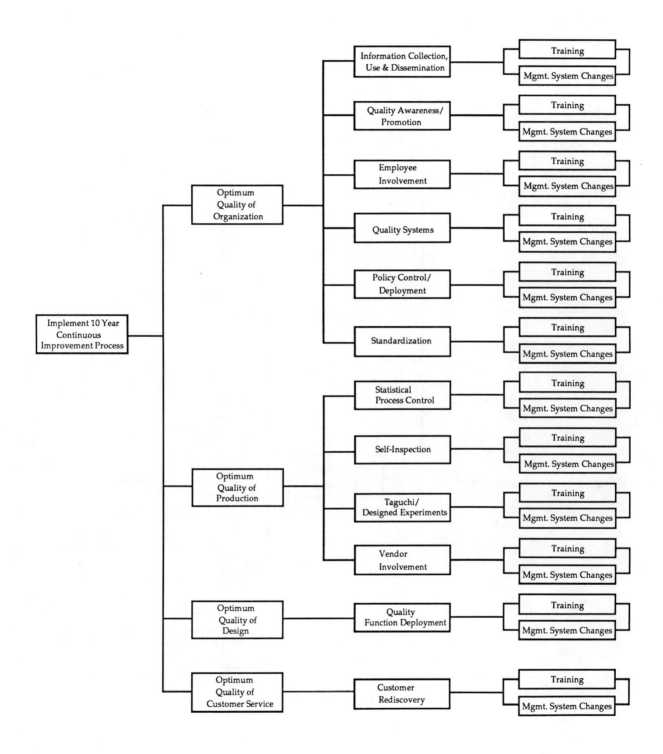

©Bob King, GOAL/QPC 7/6/89

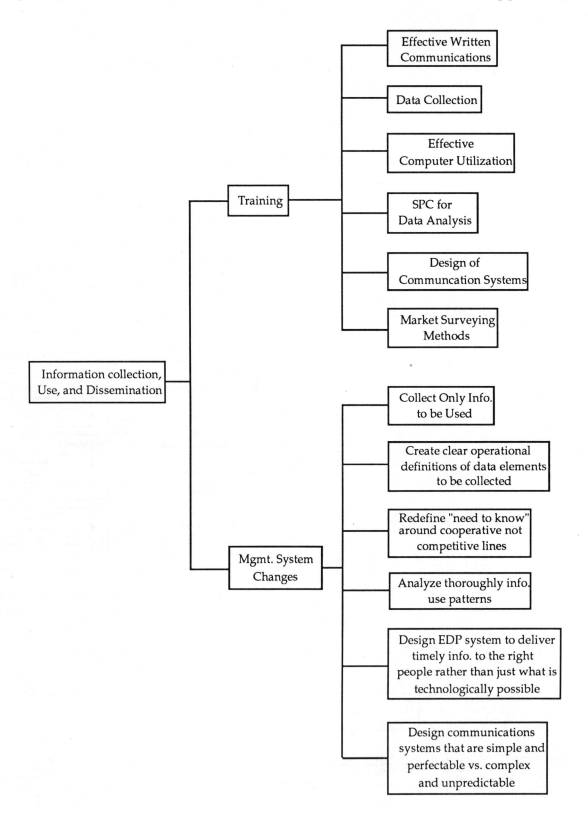

Appendix A							p. A-3

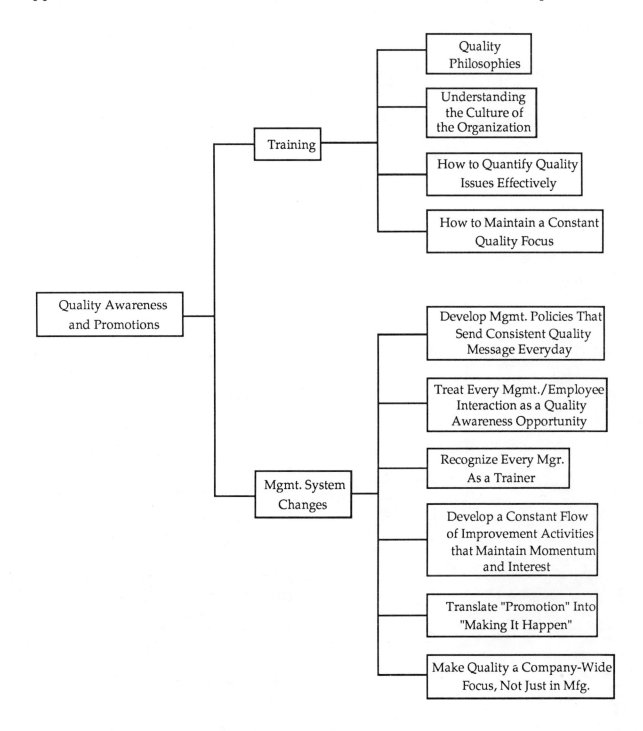

©Bob King, GOAL/QPC					7/6/89

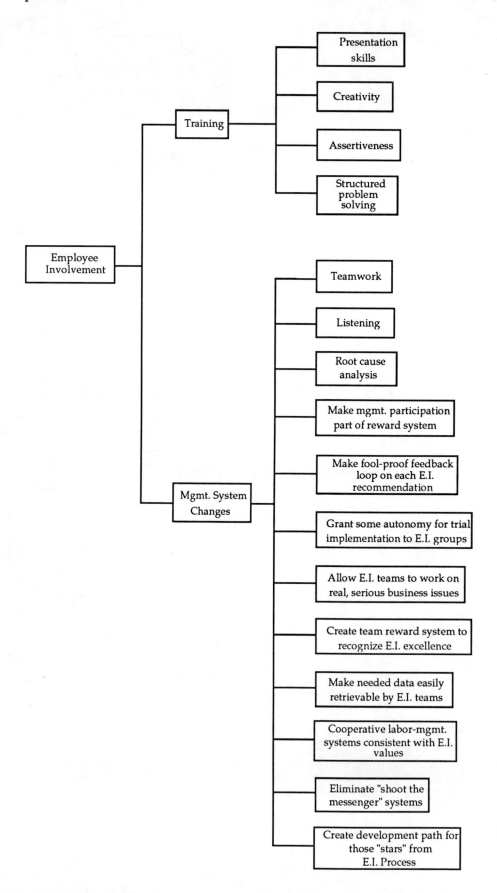

Appendix A p. A-5

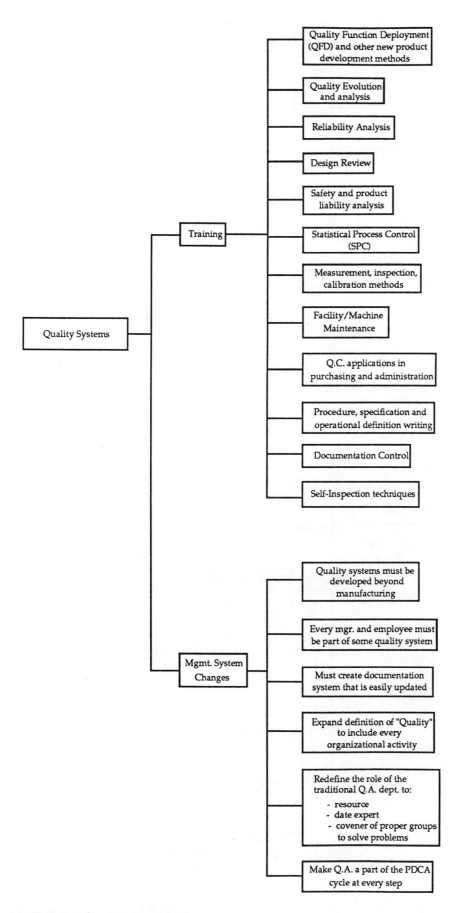

©Bob King, GOAL/QPC 7/6/89

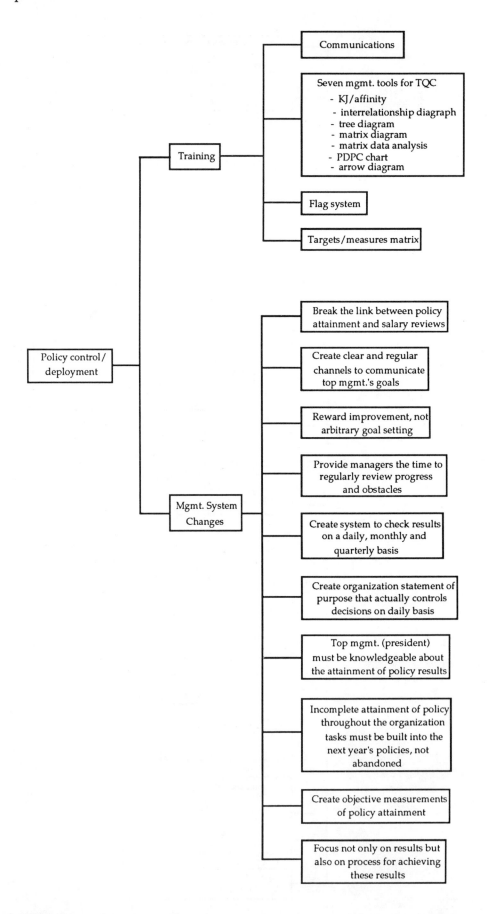

Appendix A   p. A-7

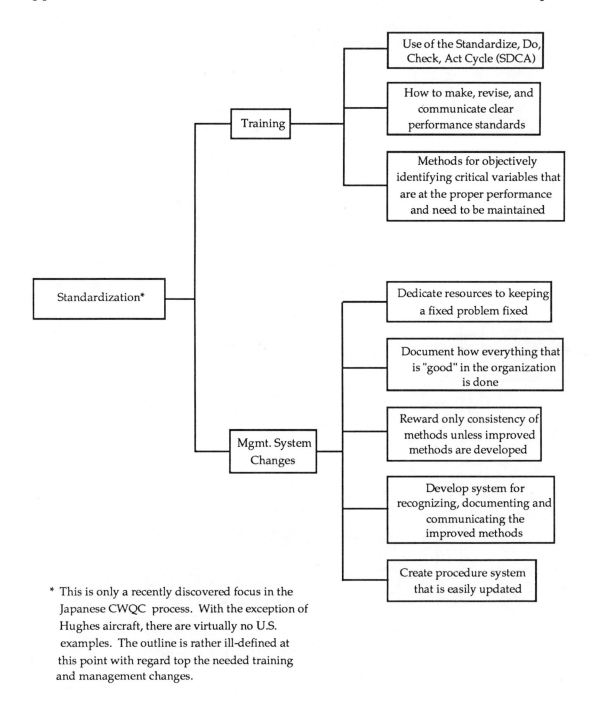

* This is only a recently discovered focus in the Japanese CWQC process. With the exception of Hughes aircraft, there are virtually no U.S. examples. The outline is rather ill-defined at this point with regard top the needed training and management changes.

©Bob King, GOAL/QPC   7/6/89

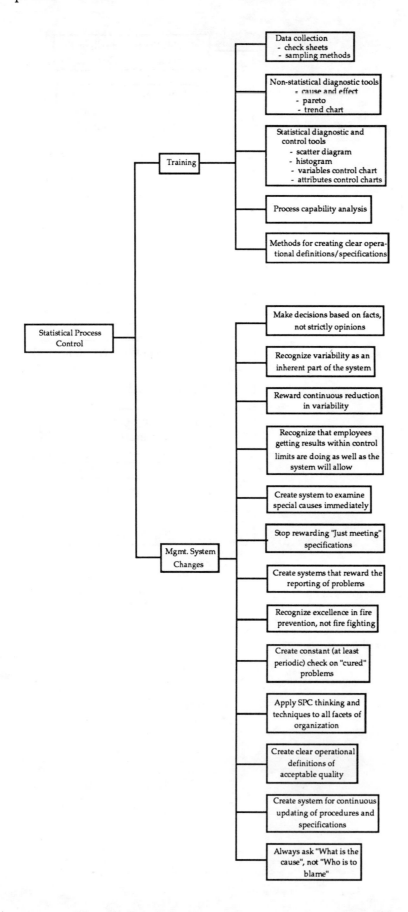

Appendix A p. A-9

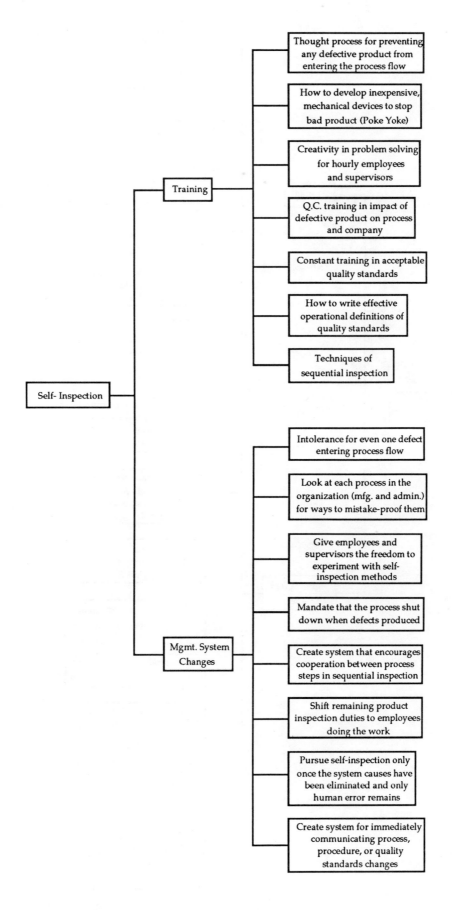

©Bob King, GOAL/QPC 7/6/89

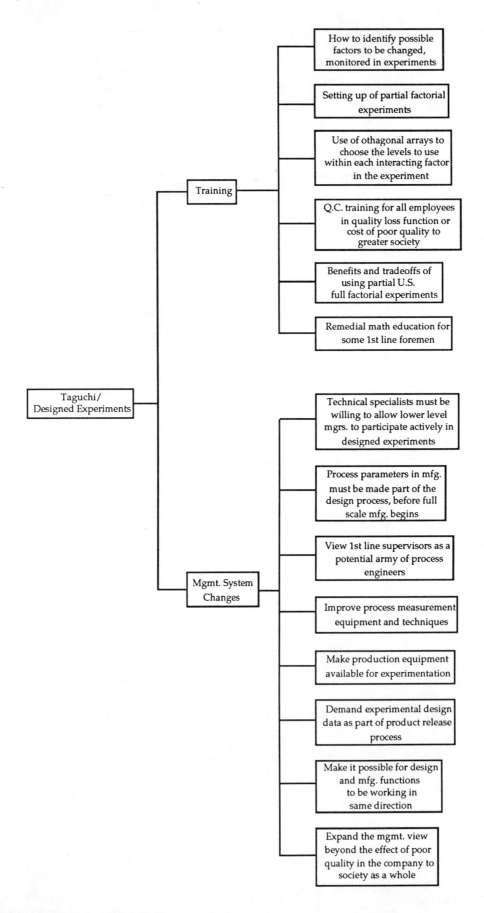

Appendix A   p. A-11

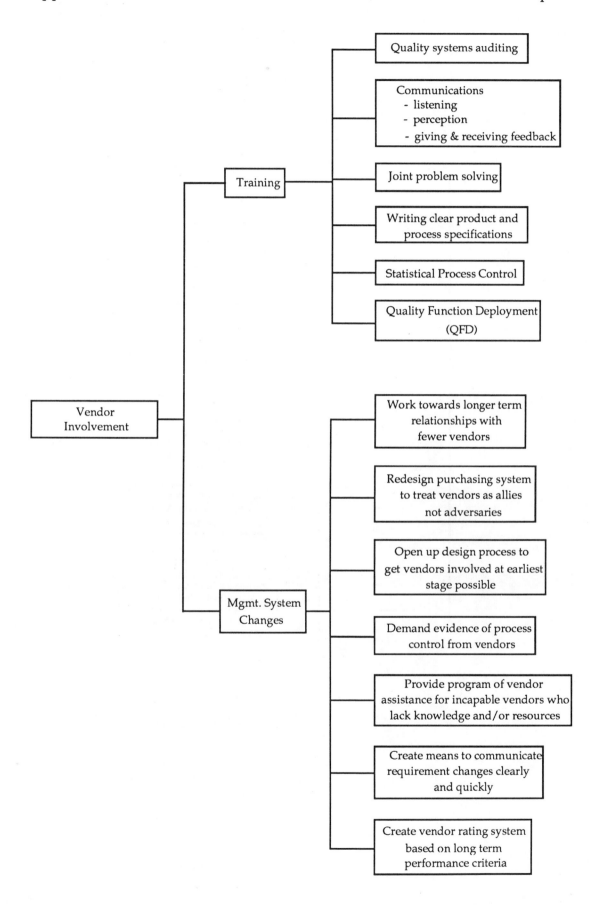

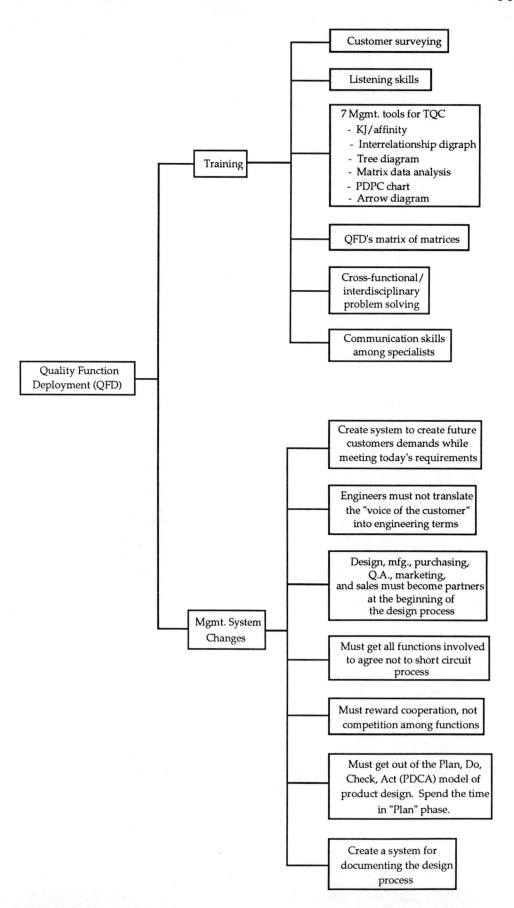

Appendix A p. A-13

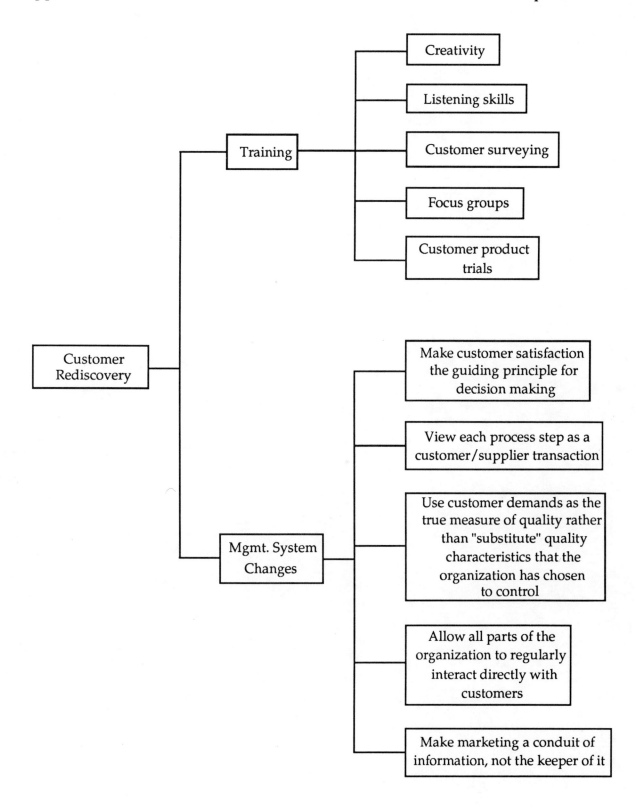

©Bob King, GOAL/QPC 7/6/89

Appendix B
Toyota Auto Body Hoshin Plan

This chapter will look at three case studies that show how policy control is being used in industry.

1. Toyota Autobody's policy control by Mr. K. Satoh, director, Toyota Autobody.

2. Total plant policy and its deployment - Sekisui Kagaku Kogyo, Sakai plant - by Mr. M. Takahashi, plant manager.

3. Policy control in the sales department - the author.

4. Takahashi's policy control received the Japan QC award in 1980 and is being introduced here for the deployment of total company policy at Toyota Autobody. This company was interested in the seven new quality tools for long time and has sent eight managers from various areas to the seven new quality tools study group meeting, from the first meeting in 1978 to the fifth in 1982, in order to actively use the tools in their company. At Toyota Autobody policy control is deployed based on the basic policy and clear management concepts. Various methods for the system such as introducing a control system per function and per step, deployment of targets, measuring by target system charts, policy deployment per function and per step by using the x-shaped matrix and others bringing major results.

2. Is a report on policy control and implementation at Sekisui, Sakai plant which received the Deming Prize in 1979. This plant is one in which I had a chance to help in its TQC promotions from 1976 until they won the Deming Prize. In the process I recommended introducing a policy control system that used the seven new quality tools and applying it to the problem solving and thus great management performance was assured as a result of the strong leadership of Mr. Takahashi, the plant manager. The results are introduced in this book. At the same time we will describe "deployment of the seven new tools in industries."

3. Is a summary of this author's experiences with TQC in the sales department based on a policy control system. Policy control in the sales department is very confidential, so that the detailed description of the situations of certain companies is difficult to publish. The author has summarized the know-how of the TQC and the sales department based on the viewpoint of the leader of the TQC. In concluding this introduction to Chapter 5, let me extend my sincerest gratitude to Messers. K. Satoh and M. Takahashi.

Appendix B: Toyota Auto Body Hoshin Plan

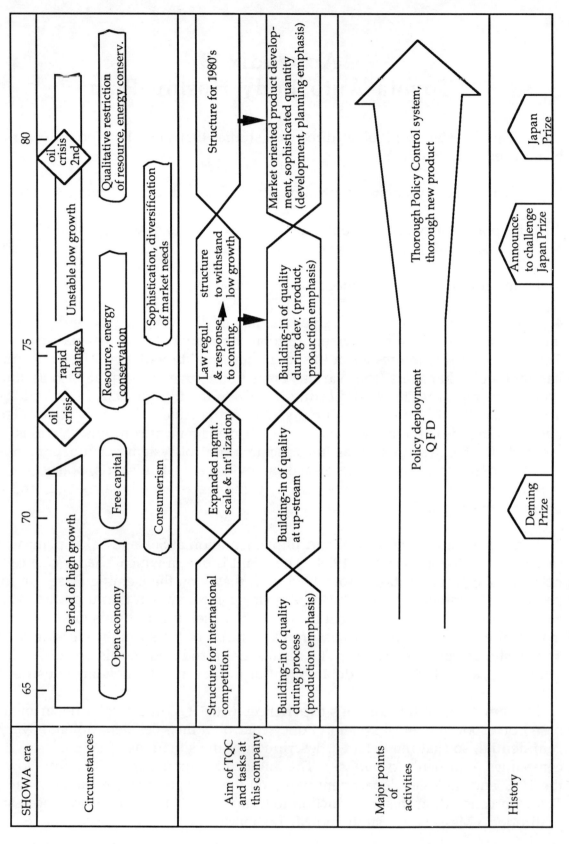

TQC Promotion History

Appendix B: Toyota Auto Body Hoshin Plan

5.1.1 The History of TQC at Toyota Autobody

Toyota Autobody began its interest in quality control and implementing it in 1954, since then it has grown to be used in the entire company since the early 1960's when TQC was introduced full-scale. Figure 5.1.1 gives a history of this period. While promoting this movement within the management environment, we have also streamlined each management task. The events can be summarized as follows.

1. Facing a growing economy 1965-1970.

2. Responding to the 1970's.

(1) Expanded scale of management and internationalization (1971-1973).

(2) Response to regulations (safety and exhaust emissions) and emergencies (oil crisis) (1974-1975).

(3) Restructuring for slow growth (1976-1977).

3. Restructuring for the 1980's (1981 to present).

(1) Responding to the opening economy (1964-1970).

In this period of free trade and free capital, Toyota Autobody has tried to introduce TQC for the purpose of "establishing a company structure that can better face international competition."

Since then, TQC has been strongly pushed for the purpose of quality assurance in the products using long term quality control in getting the participation of everyone to improve the quality of their jobs.

1. Functional control

2. New product control

3. QC circles

As a result the Deming Prize was awarded to Toyota Autobody for the attitudes and the concepts for quality control in 1970.

(2) Responding to the 1970's (1971-1979).

1. Responding to the expansion of management scales and internationalization (1971-1973).

At the period of finishing the restructuring for an open economy, Toyota came up with a series of new models and model changes due to the increase of cars in the domestic market as well as an expanding overseas market. Meanwhile, further efforts for new product control based on the valuable criticism received from the Deming Prize were used to achieve a series of targets for development.

2. Responding to regulation (safety and exhaust emissions) and emergencies (oil crisis) (1974-1975).

The oil crisis at the end of 1973 shook this company to its very roots. The strong inflationary costs had put pressure on management causing them to cancel certain model change plans and new plant constructions. On the other hand, regulations for safety and for exhaust gas emissions became more severe with the increase in consumer orientation, thus the task for Toyota Autobody during this period were two-fold.

(1) Clear regulatory hurdles for safety and exhaust emissions.

(2) Bring down costs to fight the inflation in raw material costs and labor costs. As a result of a company-wide response to these tasks, a regulation assurance system was established and large scale cost-down saved Toyota Autobody from this crisis. TQC contributed greatly to this result.

(3) Restructuring for slower growth (1976-1977)

Through this period of rapid changes, our country's economy has changed to being less stable and slower in growth.

Under these types of economic conditions, the needs of the consumer became more sophisticated and diversified. Since the market demands increased for a long lasting automobile, free of troubles, and a more personalized model for diversified concepts and values of people, Toyota Autobody revived its cancelled models and changes. Then, emphasis was placed, during this period, on the following three points.

(1) Product development in the best interests of society and consumers.

(2) A building in of quality by improving the quality of the planning.

(3) Maintaining and improving process capability in order to improve the level of quality.

In 1976, Toyota Autobody embarked upon a long term management plan which looked for an efficient progress in the above activities for restructuring based on the need to survive during the slow growth period and one of the more appropriate forms for the new coming age.

The purpose of this long term plan was to establish an overall management control system on a purpose-oriented nature. Emphasis was placed on strengthening new product development systems with a long term plan in mind. As a result enhanced development systems succeeded in a series of model changes and management control systems that were established based on this long term plan.

(3) Restructuring for the 1980's (1978 to present).

Based on changes in later situations a higher level of TQC has been pushed with an emphasis on human resources, clarifying management policies, and setting priorities for action. The concepts and procedures are explained below.

1. Recognizing the circumstances.

In the planning process of the above long term plan, Toyota Autobody decided to receive an audit for Japan QC award with the motto "to receive the most accurate evaluation by the leading authority in industry for their new development of total quality control within a certain deadline." The industrial situation for this decision was as follows. It was at the period being faced was one of:

(1) sophistication/diversification of consumer needs due to a maturity in the market, making the attractiveness of products the key for survival.

(2) A world-wide demand, an increase in competition for small cars, because of the demand for low fuel consumption due to conservation of resources and energy.

(3) Establishing reliability assurance engineering for product responsibility while smaller, lighter and more fuel efficient vehicles are needed.

(4) International inflation due to the inflation in the costs of resources and energy.

2. Emphasizing management policy.

p. B-6   Appendix B: Toyota Auto Body Hoshin Plan

Management Concept and Basic Policy

Teaching of S. Toyota

Spirit of founding of Toyota

1. Emphasize research and creation to stay ahead
2. Don't market until thorough product testing is done

Purpose of founding

Contribute to development of national economy by research/production of good products, as an exclusive plant for auto bodies

Company motto:

This company, based on its international viewpoint, research and creation, contributes to society and aims at continued expansion of its business.
Development: respect, ideas and time to stay ahead of time.
Human relationship: be pleasant and cooperative by being honest and trustworthy.
Gratitude: self-examination for progress and enjoying happiness of work

Purposes:

1. Endeavor for creative engineering development
2. Contribute to society with good products
3. Always think, respect harmony and be grateful

Basic policy:

1. Establish management to grow steadily with Toyota and contribute to rich society
2. Be No. 1 in quality and offer good products to response to requests of society and trust of customers
3. Improve management efficiency by thorough control and nationalization to strengthen company structure

©Bob King, GOAL/QPC   7/6/89

Toyota Autobody's company motto in 1953 adopted the teachings of Sakichi Toyota and the purpose of this company's founding. Together with the aims of TQC and such specific expressions, basic policies, etc. as shown in Figure 5.1.2. The basic policy which specifies the policies of the company are "to develop a company by implementing quality first." The idea is to base management on quality assurance from studies based on sales, service, assumed quantities, delivery and cost, by having priorities and using this to secure profits, which are the basic target of any new company.

Toyota Autobody decided to implement this idea as its policy and integrate it into the total company's capabilities. The emphasis here was to aim at quality assurance, arrow to the right, more cars produced by utilizing the characteristics of the company in the field of truck and commercial vehicle manufacturing. The following two were emphasized.

1. Planning for attractive products for both domestic and overseas markets.

2. Sophisticated quality to assure strength in the domestic and overseas markets.

As a result of the above implementation, one more target of QA>CR (cost reduction) happened; and this was the best way to assume the achievements of the basic targets from a long term point of view. Quality first was applied to eliminate the cost of losses due to poor quality, etc. in the line of daily QA activities and also to be more efficient in production based on VA and IE, which again results in QA>CR. This concept is shown in Figure 5.1.3.

3. The emphasis for the activities.

The above mentioned activities were deployed further in detail to achieve the management policy priorities and the emphasis there were as follows.

(1) Improved engineering development programs and product planning that is ahead of the transient society and then to market with an accurate understanding of the needs of the customer.

(2) Transmitting quality centered around quality function deployment and clarification of planned targets by use of quality deployment charts.

p. B-8   Appendix B: Toyota Auto Body Hoshin Plan

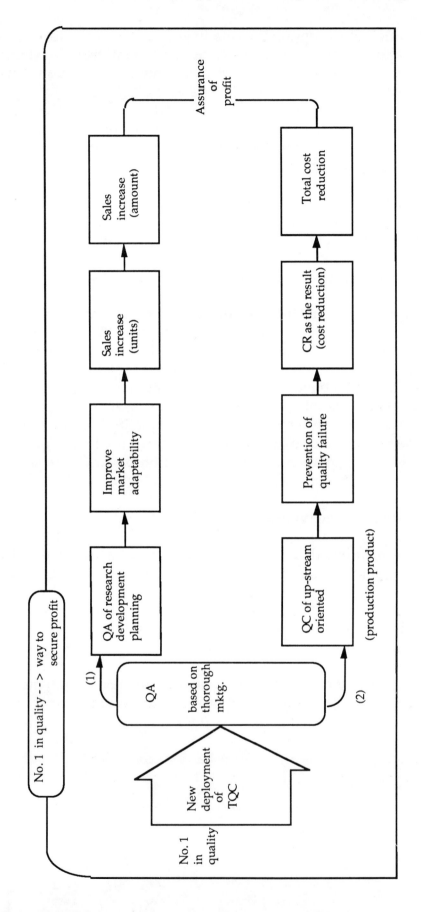

Emphasis on Management Policy Since 1978

©Bob King, GOAL/QPC   7/6/89

(3) Improving studies ahead of trouble occurrings by improving reliability forecasts, structural analyses, design audits, and facility design audits using a reliability method.

(4) Intensify evaluations and their confirmation by developing test evaluation technology.

(5) Assure process capability by improved production method deployment and process analysis.

4. Making arrangements.

A series of targets for the above activities are the core of Toyota Autobody's quality plan and are linked by use of a target system chart with profit as the final target, which is the basic target of any management. Thus, all management activities are systematized with an emphasis of the management activities at the center. They are arranged for overall management control to achieve long term and annual basic targets. The efforts of this arrangement began at the time of long term planning in 1976.

Since then it has gone through various improvements and streamlined in a form called "functional per step control system" based on the long term plan. The emphasis here is a QA system based on long term quality plans among these a strengthening of new product development systems has contributed much to the achievement of the basic targets along with improved cost control systems.

The prioritized resource allocation to the development department for assurance of this arrangement within the management organization was also an important task during this period. The details of this arrangement and its administration will be covered later.

5. Building human resources.

If your company motto says to have all employees self realize and experience joy in the job as "a job worthwhile in my life", this fosters the employees' capabilities and management's ideas has been to give them a place to utilize these abilities since the company's very founding.

The implementation of this idea is related to the building and the assurance of human resources as a management resource. This idea is the basis for the efforts in the management's training and work quality improvements for employees.

5.1.2. An Outline of Policy Control

As we talked about in the "TQC history" Toyota Autobody began planning a long term program that was suitable for the upcoming age. It began in 1976 and this has continued and efforts for improving the arrangement and the management targets for the purpose of better policy control. The major points are:

1. Function per step control system

2. Linking of target by use of a target system chart

3. Improving annual targets that are purpose oriented based on the long term perspective

Appendix B: Toyota Auto Body Hoshin Plan   p. B-11

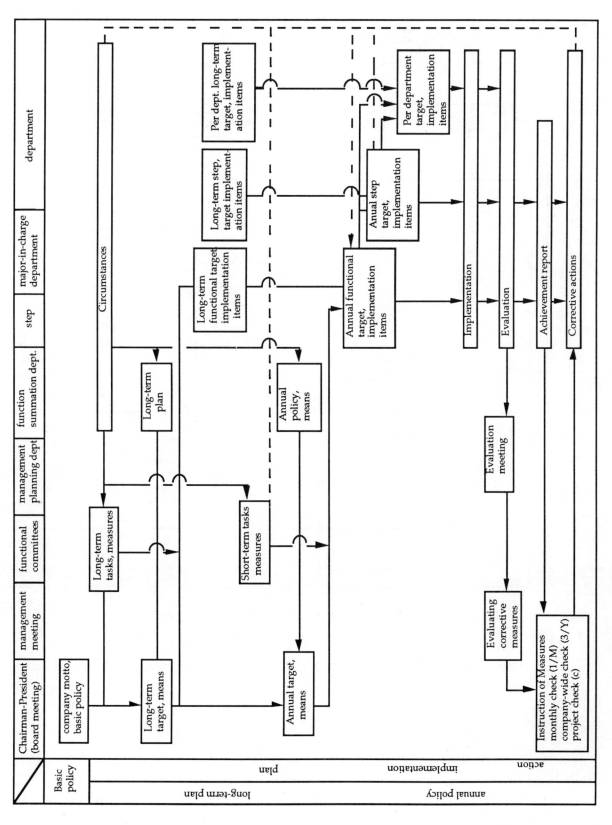

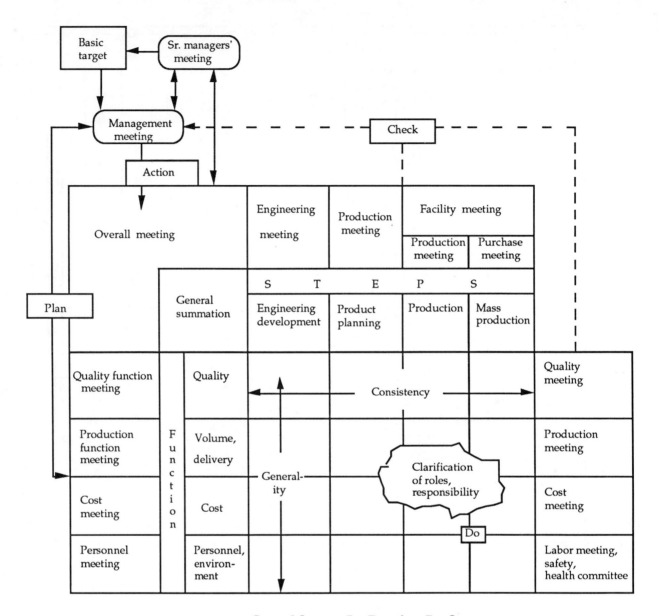

Control System, Per Function, Per Step

Toyota Autobody's policy control and the arrangement in activities are described below.

(1) The arrangement of policy control

1. Structure and flow of company policy

Toyota Autobody's policy is composed of a basic policy, long term plan, and annual policies. The long term plan and annual policies are composed of basic targets, means and function step targets for each department implementation items respectively. These are reviewed and established based

Appendix B: Toyota Auto Body Hoshin Plan

on a self examination of the previous year at the year end and any new changes in circumstance, etc. A series of plan flows, implementations, checks, actions, of company policies are shown in Figure 5.1.4.

    2. Improving the arrangement

To position management activities systematically, without omission, and to assure the achievement of the basic targets, the arrangement of the policy control was approved at Toyota Autobody by linking the targets based on the use of the target system chart, and implementation of a function per step control system.

At Toyota Autobody, as shown in Figure 5.1.5., the four stages of engineering development, product planning-product prototype, production planning, and production, along with the five functions of quality, quantity, cost, personnel environment, and general, are layered and located vertically and horizontally to form a management control system. The purpose here is to achieve a total management target by forming a network of control, systematizing management activities, while clarifying their rules and responsibilities.

(2) Functional per step control

    1. Functional control

p. B-14 Appendix B: Toyota Auto Body Hoshin Plan

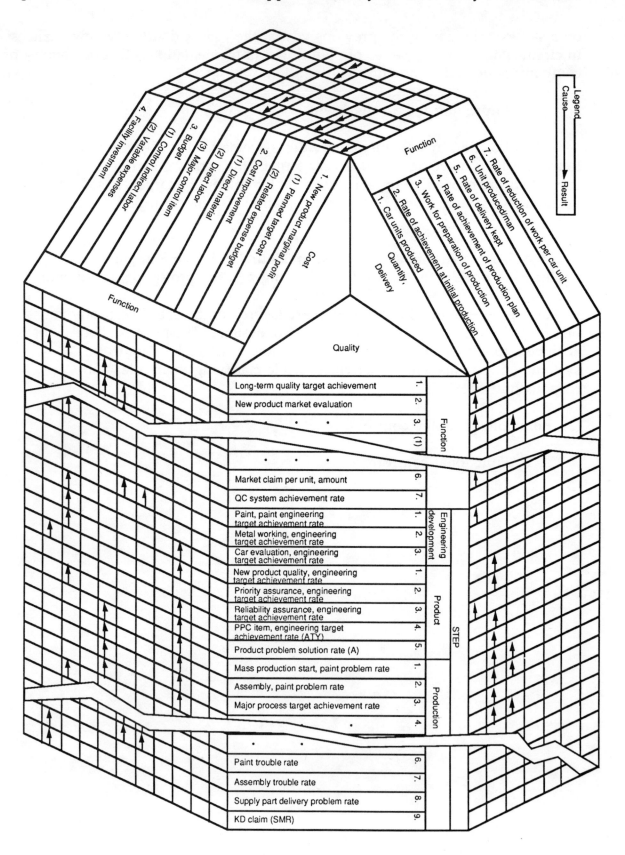

©Bob King, GOAL/QPC 7/6/89

Appendix B: Toyota Auto Body Hoshin Plan p. B-15

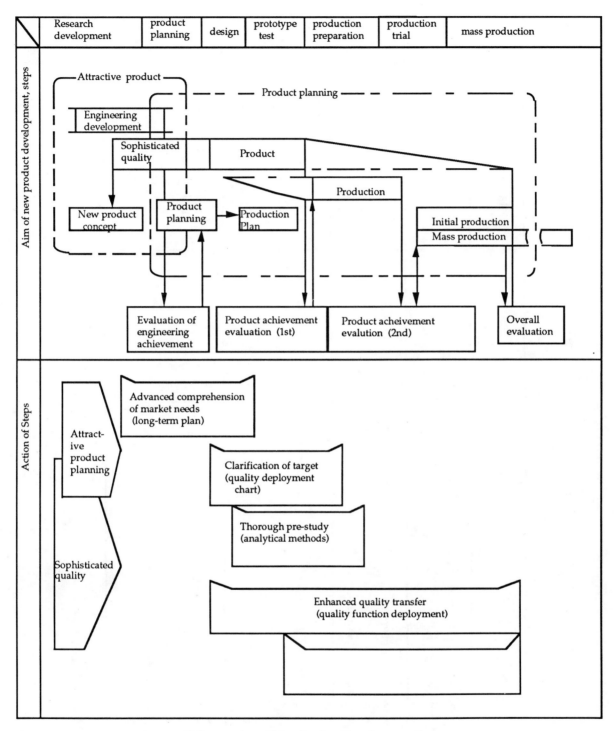

Enhancement of New Product Development System

©Bob King, GOAL/QPC 7/6/89

Major Meetings

Name of meeting	Roles	Members	Frequency	Relation to Figure 5.1.5
Sr. Managers' meeting	Conference, adjustment re: management plan, policy	Director above senior managers	2/Mo.	
Management meeting	Report, conference, adjustment re: management control	All working directors	1/Mo.	
Function meeting	Exclusive conference re: strategy, policy, target, system	Related directors	1/2 Mo.	Overall meeting, quality function meeting, etc.
Work meeting	Conference, adjustment re: deployment of target, effective promotion of target achievement	Director-in-charge related dept. head	1/Mo.	Engineering meeting, quality meeting, etc.

Appendix B: Toyota Auto Body Hoshin Plan p. B-17

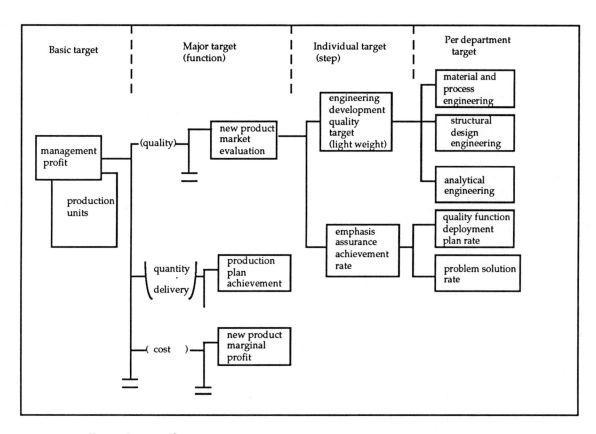

Target System Chart

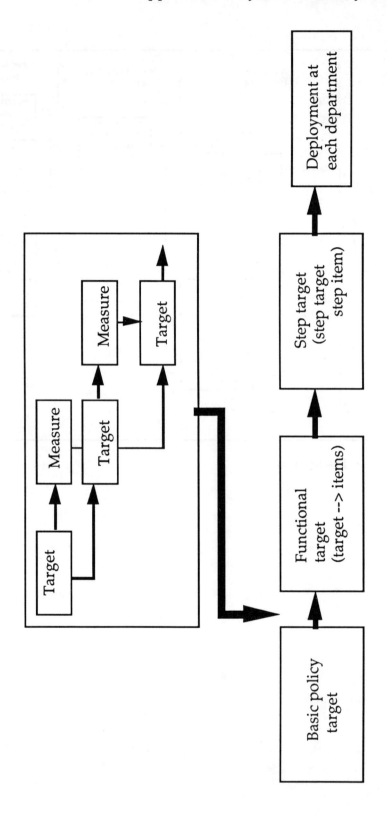

Functional control is intended to control the management activities as it relates to the separate functions of quality, quantity, delivery costs, and personnel throughout each step maintaining consistency. A broader meaning of the quality concept is to have regulated a function of the time concept while relating it to cost. In actual management activities though, it is first taken as an independent function and then it is deployed into each activity stage along with its interrelationships with others, as shown in Figure 5.1.6. To clarify the relationship with management targets the general function is separately provided for the purpose of adjustment among the functions.

2. Per step control

Per step control aims at an intensified new product development process and pushes QA arrangements one step forward as shown in Figure 5.1.7. It is to assure a sophisticated quality and an attractive product planning throughout each step. It is also intended to adjust totally the separate or contradicting means of the various functions, as shown in Figure 5.1.5.

3. Conference groups

Arranging policy control built up on functional and per step controls it provides various conference groups for the purpose of a consensus warp in implementing an efficient administration of management activities. This is seen in Figure 5.1.5. The main ones are shown in Table 5.1.1.

(3) Linking targets and means to quality control (1978-)

1. Linking of target and means

As shown in Figure 5.1.8, the linking of targets is formed to have the position targets assume higher position targets by using a target system chart to assure the achievement of the final management goals. By clarifying the mutual relationship between targets to everybody's understanding and improvement in the achievement rate of the separate targets can be obtained.

2. Deploying the target

This linking of the targets is deployed to function step in each department as a link between the target and the means as shown in Figure 5.1.9, and goes down to the very edges of management. This concept is shown in Figure 5.1.10.

The relationship of targets between function and step and between step and each department and the relationship among the implementation items are shown in Figures 5.1.11 and 5.1.12 where the conformity is perspectively established.

p. B-20 Appendix B: Toyota Auto Body Hoshin Plan

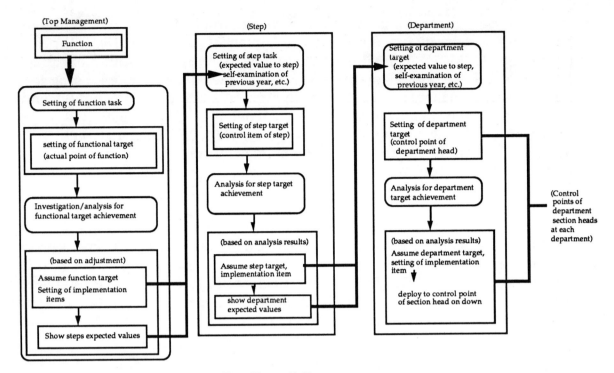

Target Measure Linking

Appendix B: Toyota Auto Body Hoshin Plan

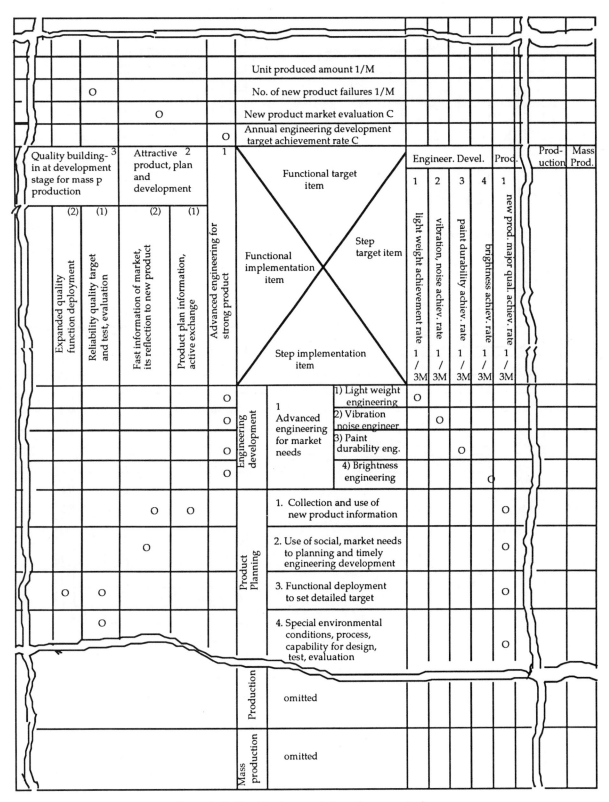

Example Policy Deployment (function ---> step)

©Bob King, GOAL/QPC 7/6/89

p. B-22 Appendix B: Toyota Auto Body Hoshin Plan

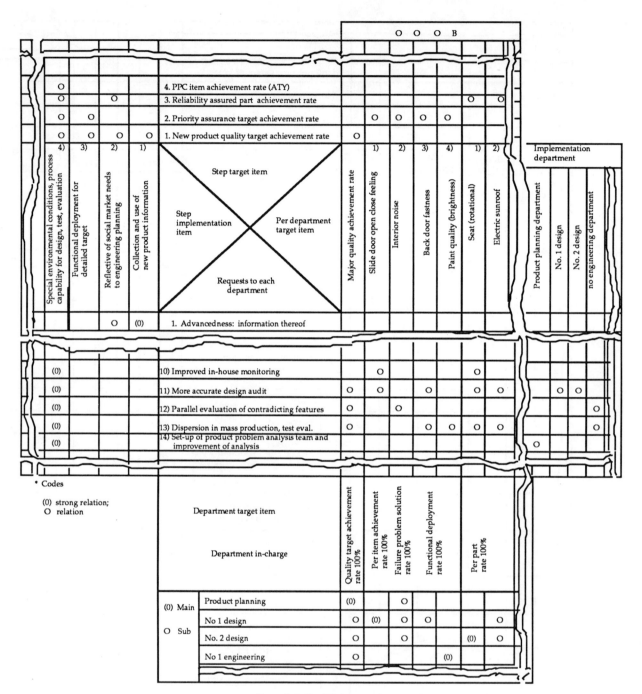

Example Policy Deployment (step —> function)

*Codes
(0) strong relation;
O relation

©Bob King, GOAL/QPC 7/6/89

Appendix B: Toyota Auto Body Hoshin Plan

	Aim	Checker	Timing
Company-wide check	- important items for management - quality of annual promotion plan - status of achievment	President & Directors	3/ Yr
Project check	Problem of new product development and confirmation of actions	President & Directors	1 month before and 4 months after line off of new product
Monthly check	All annual targets, monthly achievement status, problems and actions	President and director-in-charge of summation department	

Check of Annual Policy

(4) Setting and deploying company policy

   1. Setting the long term plan

   The long term plan is aimed at improving the quality of the plans for the annual policy an an implementation plan for the first year. Toyota Autobody reviews its long term plans every year for changes in circumstances and corrects their long term tasks and long term targets and then reconfirms the tasks and targets for the subject year. Under the long term basic targets shown by top management the long term plan is made as an overall plan mostly by the planning department with the functional integration of other departments being done to a rearrangement of a long term tasks. This long term plan is set after a conference of top management and clarifies the link between a long term basic targets>function targets>step targets>and each department's target by use of a target system chart.

   2. Planning and deploying the annual policy

   The annual policy is set based on target for the first year, shown on the long term plan, plus a self examination of the previous year and the tasks for the subject year. After the relationship between the long term target and the annual target is clarified it is deployed after linking the targets at the levels of function, step and departments which are then clarified.

(5) Implementation, check, and action in the annual policy

   The annual policy is deployed to the department level and is changed into an implementation plan to be acted upon. The status of the progress and the achievement of the annual policy are checked by the total company check project, check and monthly check for followup and for an understanding of the problems in confirmation of corrective actions and instructions which are then performed as shown in Table 5.1.2. The self examination of problems is arranged in order to properly reflect for the next year's plan or the next project. Policy revisions within the subject year may also be made.

   5.1.3 In Conclusion

   The arrangement and the administration policy control is an important core of TQC as we have introduced above. Presently functional and per step control systems are well arranged and the purpose-oriented deployment based on long term deployment of functions, step and department is being performed with good results in our opinion.

   In closing, the cautionary points for Toyota Autobody's policy can be summarized as below.

Appendix B: Toyota Auto Body Hoshin Plan p. B-25

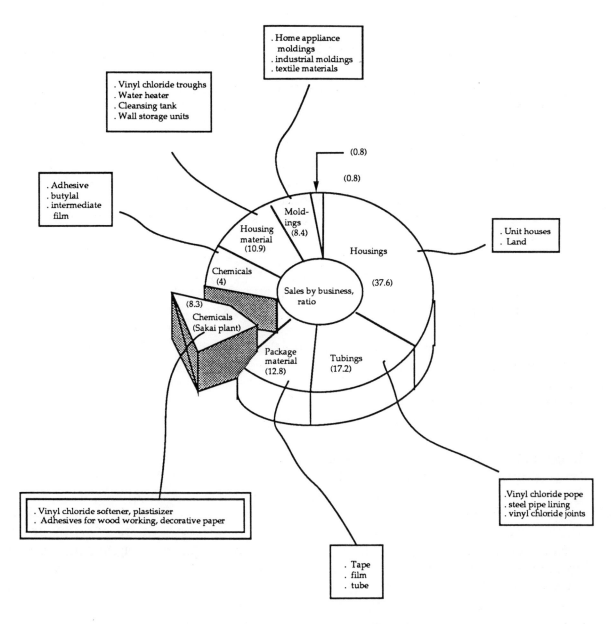

Breakdown of Products of Our Company and Its Business

1. Care must be taken to form a consensus within the company for long term plans and annual policies (step by step progress based on a long five year plan).

2. By putting basic targets for management on finance (profit) and improving employee morale. A policy control centered around a quality plan is implemented.

3. Ball catching among the individual functional integration departments, the departments in charge of the steps and the other departments is done for policy deployment. An improvement of the quality of the plan is carried out by correcting what needs to be corrected and by supplanting what needs to be supplanted.

4. In order to stabilize the interdepartmental tie-up in communication in this arrangement policy control must be systematized and standardized in order to be appropriately administered.

5.2 Total plant policy and the deployment at Sekisui Sakai plant - M. Takahashi, plant manager, Sakai plant, Sekisui

5.2.1 A company outline

The company was founded on May 3rd, 1947 and industrialized plastic modeling, which at that time was new to Japan. Later, after repeated aggressive moves into new areas, is currently involved with diversified business deployment including industrial materials, housing and chemicals.

The Sakai plant was built in May 18, 1970 as a modern chemical plant. Presently its production and sales are approximately 8% plus within the company. Figure 5.2.1 shows the business of this company, its main products, and the position of the Sakai plant.

5.2.2 The Introduction of TQC

Although the business of the company developed rapidly, the areas were quite diversified and the weakness in controls and the slowness to change began to show. To correct these conditions the Tokyo plant began its TQC and the results were recognized in the late 1970's and a Deming operation office prize was awarded. This triggered a total company adoption of TQC in 1976 with the goal of "establishing a company structure which can stand any change", and in 1979 the total company received the Deming implementation prize.

5.2.3. An outline of the Sakai plant

Index

85/15 Rule, 8-13

A

Affinity Diagram (KJ), 1-14, 4-1, 4-2

Akao, Yoji, 5-1

Arrow Diagram, 1-15, 1-16, 4-1, 4-33

Audit Tools, 1-8, 1-11

B

Brainstorming, 3-3

C

Catch Ball, 2-6

Cause and Effect Diagram, 1-13, 3-3

Check Sheet, 3-3

Continuous Improvement, 1-8, 1-17

Control Chart, 1-13, 3-1, 3-3

CPM, 1-16

Critical Path Movement (CPM), 1-16

Cross-Functional Mgmt., 1-8, 1-11

Customer-Driven Master Plan, 1-8, 1-10

D

Daily Control, 1-8, 1-10, 1-16, 1-17

Deming Cycle, 1-4

Deming, Dr. W. Edwards, 1-6

Deming's 14 Points, 8-11

Detailed Planning, 1-15, 4-1

F

Fishbone Chart, 5-1

Flag System, Ch. 6

Flow Chart, 1-13, 3-3

Ford Problem Solving Steps, 8-3, 8-7

G

Gant Chart, iv

General Planning, 1-14, 4-1

H

Histogram, 1-13, 3-3

Horizontal Teams, 1-8, 1-11

Hoshin Kanri, 1-2

Hoshin Planning, 1-8, 1-11

Hoshin Planning Name, 1-2

I

ID, 1-14, 4-1, 4-6

Information Systems, 1-8, 1-11

Inspection, 8-2

Intermediate Planning, 1-15, 4-1

Interrelationship Digraph (ID), 1-14, 4-1, 4-6

Ishikawa, Kaoru, 3-1

J

Juran, Joseph, 1-6

K

Kayaba, iii

KJ Method, 4-1, 4-2

Kogure, Masao, 1-6

Komatsu, 6-1

M

Management by Objectives (MBO), 1-3, 2-4

Management by Policy (MBP), 1-3

Maslow, 2-13

Matrix Data Analysis, 1-15, 4-1, 4-26

Matrix Diagram, 1-15, 4-1, 4-15

MBO, 1-3, 2-4

Mizuno, Shigeru, 3-1

N

Nayatani, Y., 1-14

O

Organizational Tools, 1-10

P

Pareto Chart, 1-13, 3-3

PDCA cycle, 1-4, 1-17, 1-18

PDPC, 1-15, 1-16, 4-1, 4-28

Performance Appraisals, 2-6, 8-13

PERT, 1-16

Plan-Do-Check-Act Cycle (PDCA), 1-4, 1-17, 1-18

Policy Deployment, 1-2

Problem Solving Tools, Ch. 3

Process Decision Program Chart (PDPC), 1-15, 1-16, 4-1, 4-28

Process Management, Ch. 8

Program Evaluation and Review Technique (PERT), 1-16

Q

QCDP, 1-8, 1-11, 3-2

QFD, 1-8, 1-11

Quality Circles, 1-8, 1-10

Quality Function Deployment (QFD), 1-8, 1-11

R

Relations Diagram, 1-14

Run Chart, 1-13, 3-3

S

Scatter Diagram, 1-13, 3-3

SDCA Cycle, 1-17, 1-18

Self-Diagnosis, Ch. 9

Seven Management Tools, 1-8, 1-14, Ch. 4

Seven QC Tools, 1-13, 3-2

Shewhart Cycle, 1-4

SMTQC, 1-7

Standardize-Do-Check-Act Cycle (SDCA), 1-17, 1-18

Standardization, 1-8, 1-10

Statistical Methods/Tools, 1-8, 1-10, 1-13

Strategic Management Total Quality Control (SMTQC), 1-7

System Chart, v

System Flow (or Tree Diagram), 4-10

T

Taguchi, 3-2

Target/Means Matrix, Ch. 5

Taylor, Frederick, 1-3

Total Quality Control (TQC), Ch. 1

Total Quality Management (TQM), Ch. 1

TQC, Ch. 1

TQC Wheel, 1-8

TQM, Ch. 1

Tree Diagram, 1-15, 4-1, 4-10

U

United Total Quality Control (UTQC), 1-7

UTQC, 1-7

V

Vertical Teams, 1-8

W

Work Groups, 1-8, 1-10

Y

Yokagawa Hewlett-Packard, i

Glossary

<u>85/15 Rule</u>: Deming's belief that management has 85% control over the system and workers have 15% control; the bulk of the causes of low quality and low productivity belong to the system and thus lie beyond the power of the work force.

<u>Affinity Diagram</u>: (Also known as KJ Diagram) One of seven management tools that assists general planning. It makes sense out of disparate language information by placing it on cards and grouping the cards that go together in some creative way. "Header" cards are used to summarize each group of cards.

<u>Arrow Diagram</u>: One of the seven management tools which, like the PERT (Program Evaluation Review Technique) and CPM chart (Critical Path Method), assists detailed planning by plotting the sequence of steps for doing parts of a job, indicating the jobs that can be done simultaneously, and the "critical" or longest path from beginning to end.

<u>Audit Tools</u>: Part of the TQC wheel. Used to monitor improvement activities at each level of the organization to assure that each employee reaches his or her full potential.

<u>Catch Ball</u>: A term that refers to the fact that communication up, down, and horizontally across the organization must sometimes go from person to person several times to be clearly understood.

<u>Cause and Effect Diagram</u>: A problem solving statistical tool which indicates effects and causes and how they interrelate.

<u>Continuous Improvement</u>: A system in which individuals in an organization look for ways to do things better, usually based on understanding and control of variation.

<u>Control Chart</u>: A problem solving statistical tool which indicates whether the system is in or out of control and whether the problem is a result of special causes or common system problems.

<u>Control Item</u>: Something measured to determine to what extent a target or means (measure) is met.

<u>Critical Path Movement (CPM)</u>:

<u>Cross-Functional Management</u>: Part of the TQC wheel. Cross-Functional

Management is used so that that all aspects of the organization are well-managed and have consistent, integrated quality efforts pertaining to scheduling, delivery, plans, etc.

Customer-Driven Master Plan: The heart of the TQC wheel. This plan ensures that the organization will know what customers will want 5-10 years from now and that the company knows what it will do to exceed expectations.

Daily Control: The system by which each worker identifies simply and clearly what he must do to fulfill his job function in a way that will enable the organization to run smoothly. Also the system by which these required actions are monitored by the employee himself. Part of the TQC wheel.

Deming Cycle: See PDCA Cycle.

Deming's 14 Points: These set the stage for successful process management.

Flag System: A method of displaying the interrelationships of different organizational run charts in such a way that planned and actual results are clear as well as relationships between various units. The combination of lines and run charts has the appearance of flags, hence the name.

Flow Chart: A problem solving statistical tool which shows the way things go through the organization, the way they should go, and the difference.

Histogram: A chart which takes measurement data (e.g., temperature, dimensions) and displays its distribution. A histogram reveals the amount of variation that any process has within it.

Horizontal Teams: Part of the TQC wheel. Teams of people who communicate about quality, cost, delivery, productivity, etc. so that all aspects of the organization are in sync.

Hoshin Kanri: The Japanese name for hoshin planning. In Japanese, hoshin kanri means "shining metal" and "pointing direction".

Hoshin Planning: One of three major systems that make TQC possible. Hoshin planning helps to control the direction of the company by orchestrating change within a company. This system includes tools for continuous improvement, breakthroughs, and implementation. The key to hoshin planning is that it brings the total organization into the strategic planning process, both top-down and bottom-up. It ensures that the direction, goals, and objectives of the company are rationally developed, well defined, clearly communicated, monitored, and adapted based on system feedback.

Information Systems: Part of the TQC wheel. Information systems allow information to flow smoothly and concisely to all the people who need it.

Interrelationship Digraph (ID): (Also called Relations Diagram) One of the seven management tools which assists general planning. The ID shows with arrows the cause and effect relationship between items. Important items are recognized by the high number of arrows going in and coming out. Items with arrows only going out are usually good places to initiate action.

KJ Method: See Affinity Diagram.

Matrix Data Analysis: One of the seven management tools which assists intermediate planning. Matrix Data Analysis shows relationships between individual items and groups of items by plotting information on x- and y-axes.

Matrix Diagram: One of the seven management tools which assists intermediate planning. The Matrix Diagram compares one set of items against another set and identifies the strength of their relationship.

MBP: (Management by Policy) Another name for hoshin planning (used in Japan).

MBO: (Management by Objectives) A system of planning which focuses on the targets that must be met to enhance the profit of the organization.

Means: (or Measure) A way to accomplish a target.

Monthly Audit: The self-evaluation of performance against targets; an examination of things that helped or hindered performance in meeting the targets, and the corrective actions that will be taken.

Objective: The means to meet a one year plan at the top of the organization or to support the base strategy at a lower level of the organization.

One year plan: A statement of objectives of an organizational event for a year.

Organizational Tools: These provide a team approach in which people get together to work on problems and also to get better at what they are doing. Organizational tools include work groups and quality circles.

Pareto Chart: A vertical bar graph showing the bars in order of size from left to right. Named after the 19th century economist Wilfredo Pareto (who discerned that wealth was not evenly distributed). Helps focus work on the vital few problems rather than the trivial many.

PERT Chart: (Program Evaluation and Review Technique) This chart aids in the reduction of over-all project time by showing which things can be done simultaneously and enabling a reduction of delays between things which are done sequentially.

PDCA Cycle: (Plan-Do-Check-Act) The plan-do-check-act system, sometimes referred to as the Deming or Shewhart cycle, is the scientific methodology in which improvements are planned, tried, and checked to see if they deserve to be implemented or abandoned.

Plan: The means to achieve a target.

Policy: The company objectives that are to be achieved through the cooperation of all levels of managers and employees. A policy consists of targets, plans, and target values.

Policy Deployment: One English translation for hoshin kanri. (Others are management by policy and hoshin planning.) Policy deployment orchestrates continuous improvement in a way that fosters individual initiative and alignment.

Problem Solving Tools: These tools find the root causes of problems. They are tools for thinking about problems, managing by facts, and documenting hunches. The tools include: check sheet, line chart, pareto chart, flow chart, histogram, control chart, and scatter diagram. In Japan, these are called the Seven QC Tools.

Process Decision Program Chart (PDPC): A tool for detailed planning that identifies the various things that can go wrong in a plan and also identifies the countermeasures or contingencies.

Process Management: This involves focusing on the process rather than the results. A variety of tools may be used for process management, including the Seven QC Tools.

Run Chart: This chart visually represents data. It is used to monitor a system to see whether or not the long range average is changing.

QCDP: Quality, Cost, Delivery, Profit, or Product are the most common areas in which cross-functional management takes place. QCDP teams are formed to ensure that quality, cost, efficiency, services, and profit are managed on a consistently high level throughout all levels of the organization. Part of the TQC wheel.

Quality Circles: Part of the TQC wheel. Quality circles are an organizational tool that provide a team approach in which people get together to work on problems and to improve productivity.

Quality Function Deployment (QFD): (Also called Quality Factor Development) A system that identifies the needs of the customer and gets that information to all the right people so that the organization can effectively exceed competition in meeting the customers' most important needs, thereby increasing market share. Part of the TQC wheel.

Quality Management: The systems, organizations, and tools which make it possible to plan, manufacture, and serve a quality product or service.

Relations Diagram: See Interrelationship Digraph.

Run Chart: A statistical problem solving tool which shows whether key indicators are going up or down and whether that's good or bad.

Scatter Diagram: One of the Seven QC Tools. The Scatter Diagram shows the relationship between two variables.

SDCA Cycle: (Set standard, Do it, Check it, Activate or Adjust it) This is the system by which standards procedures are implemented or changed.

Self-Diagnosis: As a basis for continuous improvement, each manager uses problem solving activity to see why he or she is succeeding or failing to meet targets. This diagnosis should focus on identifying personal and organizational obstacles to planned performance and on the development of alternate approaches based on this new information.

Seven New Tools for Management and Planning: (Also called the Seven Management Tools or Seven New Tools) These are affinity chart and relations diagram (interrelationship digraph) for general planning; tree (systems), matrix and matrix data analysis for intermediate planning; and arrow and process decision program chart for detailed planning. Part of the TQC wheel.

Seven QC Tools: Problem solving statistical tools needed for the customer-driven master plan. They are: cause and effect diagram, flow chart, pareto chart, run chart, histogram, control chart, and scatter diagram.

Shewhart Cycle: See PDCA Cycle.

Standardization: The system of documenting and updating procedures to make sure everyone knows clearly and simply what is expected of them (measured by daily control). Part of the TQC wheel.

Statistical Methods/Tools: Part of the TQC wheel. Statistical methods allow employees to manage by facts and analyze problems through understanding variability and data. The Seven QC Tools are examples of statistical tools.

Statistical Quality Control (SQC): The use of knowledge and control of variation to improve and maintain the organization's processes.

Storyboarding: A system developed and popularized by Walt Disney in which characters, plots, locations, etc. were put on cards and reorganized to create new story lines. It is now used to describe any use of cards to arrange information.

Strategic Management Total Quality Control (SMTQC): SMTQC is related to UTQC and plays a part in the development of the optimal TQC system in a multi-national company, in which there are similarities in the quality program corporate wide, but the quality program is customized for each company. In his 1988 book, Masao Kogure predicts that SMTQC and UTQC will be the major growth areas of TQC in the future.

Strategy: A means to meet an objective.

System Flow/Tree Diagram: See Tree Diagram.

Target: The desired goal that serves as a yardstick for evaluating the degree to which a "policy" is achieved. It is controlled by a "control point", "control item", or "target item".

Target/Means Matrix: Developed by Yoji Akao to show the relationship between targets and means and to identify control items and control methods.

Target Value: Normally a numerical definition of successful target attainment. (It is not always possible to have a numerical target). You must never separate the target from the plan.

Total Quality Control (TQC): The management system in which all employees in all departments improve or maintain quality, cost, yield, procedures, and systems to give customers a product or service which is most economical, useful, and of the best quality.

Total Quality Management (TQM): See TQC.

Tree Diagram: One of the seven management tools that assists intermediate planning. The Tree Diagram systematically breaks down plans into component parts. It systematically maps out the full range of tasks/methods needed to achieve a goal. It can either be used as a cause-finding problem solver or a task-generating planning tool.

TQC Wheel: A schematic that summarizes the systems which comprise Total Quality Control and the interrelations of those systems.

United Total Quality Control (UTQC): UTQC is related to SMTQC and plays a part in the development of the optimal TQC system in a multi-national company, in which there are similarities in the quality program corporate wide, but the quality program is customized for each company. In his 1988 book, Masao Kogure predicts that UTQC and SMTQC will be the major growth areas of TQC in the future.

Vertical Teams: Vertical teams are groups of people who come together to meet problems or challenges. These teams are made up of the most appropriate people for the issue, regardless of their levels or jobs within the organization. Vertical teams are part of the TQC wheel.

<u>Vision</u>: A five to ten year plan of direction that is based on a careful assessment of the most important directions for the organization.

<u>Work Groups</u>: Work groups are an organizational tool. They provide a team approach in which people get together to work on problems or improve productivity. Work groups are part of the TQC wheel.

Hoshin Planning   Conclusions

1. Easing from the existing system to hoshin planning

The transition from the existing system to hoshin planning depends on the existing system. An important issue is the organization's current position on the maturity pyramid.

Some of the following kinds of questions may help determine the organization's position.

		strongly agree			strongly disagree	
1.	Targets are announced with no plan of how to reach them.	5	4	3	2	1
2.	Each year a new plan is developed from a clean sheet of paper.	5	4	3	2	1
3.	If a target is not met, a major effort is made to determine why it was not met.	5	4	3	2	1
4.	Appropriate data is gathered and analyzed so that people know what is going on.	5	4	3	2	1
5.	People have a clear idea of what to expect of other people and what other people expect of them.	5	4	3	2	1

6.	Basic problem solving tools are regularly used to decide how the organization can be improved.	5	4	3	2	1
7.	Each manager has two or three written goals and reviews progress on them each month.	5	4	3	2	1
8.	Each manager evaluates his/her own missed goals to understand problems and make adjustments.	5	4	3	2	1
9.	Each manager uses learnings about past problems to help set better goals for the future.	5	4	3	2	1
10.	Managers are generally aware of the key goals of others in the organization.	5	4	3	2	1
11.	Managers do not consciously select goals which will adversely effect other managers in the organization.	5	4	3	2	1
12.	Managers from various departments get together every one or two months to coordinate efforts on quality, cost, and delivery of products and services.	5	4	3	2	1
13.	The top two or three priorities of the organization are posted and described thoroughly in the organization's magazine, etc. Everybody knows the strategy to accomplish.	5	4	3	2	1
14.	Everybody knows the organization's top priorities.	5	4	3	2	1
15.	Every person has his or her own top two or three priorities which are tied to those of the organization.	5	4	3	2	1

Bibliography

Akao, Yoji, editor. <u>Practical Applications of Management by Policy</u>, Japan Standards Association. Tokyo 1988 (Japanese) 245 pgs.

*Imai, Masaaki <u>Kaizen and the Key to Japan's Competitive Success</u> Random House Business Division, New York 1986 pp. 94, 125, 142-145

*Ishikawa, Kaouro <u>What is Total Quality Control? The Japanese Way</u> Prentice Hall, Inc. Englewood Cliffs, N.J. 1985 pp. 59-61, 125-126

*Mizuno, Shigeru <u>Company-Wide Total Quality Control</u> Nordica International Limited. Hong Kong, 1988 pp. 97-106

Nyatani, Yoshimobu <u>Using the Seven New Tools for Improved Policy Management for Total Quality Control</u>, JUSE Press. Tokyo 1982, (Japanese) 240 pgs.

*available through GOAL/QPC, 13 Branch Street, Methuen, MA 01844
(508) 685-3900